Signals and Systems for Electrical Engineers I

by Dennis M. Sullivan

Copyright © 2018 Dennis M. Sullivan

All rights reserved.

ISBN-10: 1722709459
ISBN-13: 978-1727094541

Dedication

This book is dedicated to anyone tough enough to study electrical engineering.

Preface

This book grew out of the course notes from an undergraduate class taught over a ten year period. The goal of the class was to teach electrical engineering students the Laplace, Fourier, and Z transforms, as well as related topics that are the cornerstones of theory in electrical engineering.

The five chapters in this book are used for a one semester class. The class is divided into four sections: The first two chapters form the first section, and then the next three sections cover chapters three through five. An exam is given at the end of each section and a comprehensive exam is given at the end of the semester.

The order in which the material is presented is based on what has seemed most palatable to the students. For instance, most books teach Fourier transforms before Laplace transforms. But my students have had some experience with Laplace transforms before the class, and find it easier to deal with Laplace transforms before going to Fourier theory and Z transforms.

Even though this is fundamentally a mathematics class for electrical engineering, the approach used is to describe the problem and then introduce the theory needed to solve it. I have avoided presenting theorems and proofs without motivation. However, I have made a point of including theory that will allow students to transition to more rigorous mathematics as they proceed in their education.

This class represents the elegant mathematics that was my main motive for studying electrical engineering. I hope all electrical engineering students will appreciate this and learn to utilize it in their careers.

Dennis M. Sullivan
Moscow, Idaho
A solution manual is available to teachers by contacting the author.

Acknowledgements

The author is grateful to Ms. Judy Lalonde for her excellent editing, to Ms. Stephanie Bunney for crucial word processing help, and to John Jackshaw for his endless computer support.

Contents

Dedication   iii

Preface   iv

1. Introduction   2

    1.1 Complex Numbers
        1.1.1 Basic Properties   2
        1.1.2 Euler's Equations   5
        1.1.3 Phasor Notation   12
        1.1.4 Phasors to Solve Differential Equations   14
        1.1.5 Summary of Phasors   19
    1.2 Signals   21
        1.2.1 Introduction to Signals   21
        1.2.2 Special Signals   21

2. Systems and Convolution   32

    2.1 Linearity and Time Invariance   32
    2.2 System Analysis   33
        2.2.1 The Response to Initial Conditions   34
        2.2.2 The Impulse Response and Convolution   35
        2.2.3 System Analysis Using the Impulse Response   41
    2.3 Convolution Properties   48
        2.3.1 Special Convolutions   52
        2.3.2 Graphical Convolution   52

3. Laplace Transforms   56

    3.1 Introduction to the Laplace Transform   56
        3.1.1 Partial Fraction Expansion   62
    3.2 Theorems and Properties of the Laplace Transform   66

3.2.1 Stability and the Laplace Transform   72
3.3 Solving Differential Equations with Laplace Transforms   75
    3.3.1 Partial Fraction Expansion for Multiple Roots   82
3.4 The Inverse of Laplace Transforms with Multiple Roots   84
    3.4.1 The Inverse of a Complex Conjugate Pair   85
    3.4.2 The Cosine Method   87
    3.4.3 The Sine Method   88
3.5 Block Diagrams   96
    3.5.1 Block Diagram Analysis   97
    3.5.2 Block Diagrams Using Error Functions   103

4. Fourier Theory   118

4.1 Introduction to Fourier Transforms   118
    4.1.1 Fourier Transforms and the Delta Function   118
    4.1.2. Properties of Fourier Transforms   119
    4.1.3 The Gaussian Function   137
    4.1.4 Using Fourier Transforms to Solve Integrals   139
    4.1.5 Review of Sections 4.1   139
4.2 Fourier Series   140
    4.2.1 Introduction to Inner Products   140
    4.2.2 The Basis of the Fourier Series   141
    4.2.3 The Complex Fourier Series   144
    4.2.4 The Fourier Series Using Fourier Transforms   147
4.3 Fourier Transforms of Fourier Series   153
4.4 Bode Plots   158
    4.4.1 The Unit of Decibels   158
    4.4.2 Introduction to Bode Plots   159
    4.4.3 Drawing Bode Plots   168
    4.4.4 Determining the Transfer Function from a Bode Plot   173
4.5 Block Diagrams in the Frequency Domain   175

5. Z Transforms   190

5.1 Discrete Signals   190

5.2 Introduction to the Z Transform	196
5.2.1 Properties of Z Transforms	197
5.3 Inverse Z Transforms	208
5.3.1 Partial Fraction Expansion of Multiple Poles	215
5.4 Inverse Z transforms of Complex Poles	219
5.4.1 The Cosine Method	219
5.4.2 The Sine Method	220
5.4.3 Summary	222
5.5 Solving Difference Equations Using Z Transforms	227
5.6 Stability Analysis in the Z Domain	236
5.7 Block Diagrams in the Z Domain	238

Chapter 1. Introduction

1.1 Complex Numbers

By now we have all becomes very comfortable working with real numbers. We know their properties:

1. Addition $\qquad a+b=c$
2. Multiplication $\qquad a \cdot b = c$
3. Powers $\qquad a^b = c$.

We know their inverses:

1. Subtraction $\qquad c - b = a$
2. Division $\qquad \dfrac{c}{b} = a$
3.a Logarithms $\qquad b = \log_a c$
3.b Roots $\qquad a = \sqrt[b]{c}$

1.1.1 Basic Properties of Complex Numbers

We are also used to seeing numbers plotted on a line. We start with the integers, 1,2,3, that form the evenly space intervals in the positive direction. We can perform simple arithmetic operations like 2 + 3 = 5 that lead us to the positive integer values on the line shown in Fig. 1.1. If we started looking at subtraction, 3 – 5 = -2, then we arrive at the negative integer values. Similarly, when we start thinking about division, we get rational numbers, like 3/5 = 0.6 that fit in the spaces between the lines. We can even think about irrational numbers like $\sqrt{2} = 1.4142...$ that we can only approximate, but can still picture them as having a place somewhere on our line.

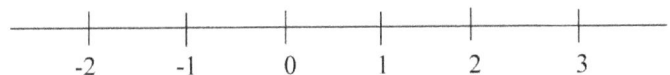

Figure 1.1. A line containing real numbers stretching from −∞ to ∞.

What happens when we are faced with an equation in the form,

$$x^2 = -1 ? \qquad (1.1)$$

The solution must be a number which can be squared to give minus one. We will give it the name j,

$$x = j = \sqrt{-1}. \qquad (1.2\ a)$$

Notice that Eq. (1.1) has another solution,

$$x = -j = -\sqrt{-1}. \qquad (1.2\ b)$$

We might be faced with an equation like Eq. (1.3)

$$x^2 + 2x + 2 = 0 \qquad (1.3)$$

When we solve it, we see that it has actually has two solutions,

$$x = \frac{-2 \pm \sqrt{2^2 - 4 \cdot 2}}{2} = -1 \pm \sqrt{-1} = -1 \pm j. \qquad (1.4).$$

We will have to modify Fig. 1.1 to accommodate these new numbers. This is shown in Fig. 1.2

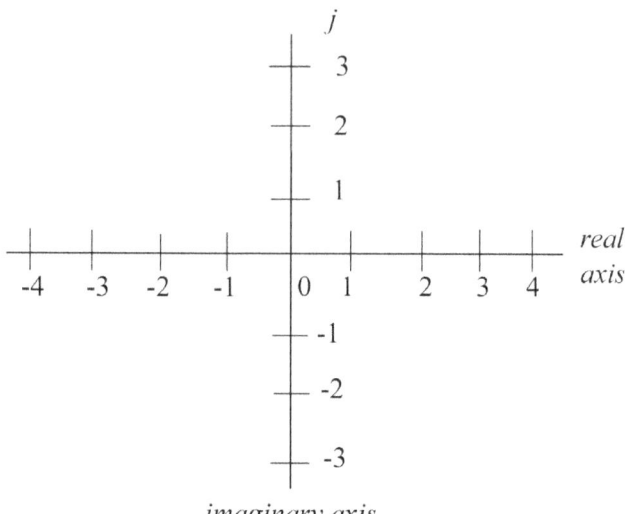

Figure 1.2. The complex plane containing real numbers in the horizontal direction and imaginary numbers in the vertical direction.

We can construct *complex numbers* in the form

$$x = p + jq, \qquad (1.5\text{ a})$$

where p and q are real numbers. For every complex number like Eq. (1.5 a), there is a number called the *complex conjugate*,

$$x^* = p - jq. \qquad (1.5\text{ b})$$

Once we have decided we are stuck with these weird numbers in Eq. (1.5), we must wonder which if any, of the above properties hold? For instance, can we add the following two complex numbers together?

$$x_1 = a_1 + jb_1, \qquad (1.6\text{ a})$$

$$x_2 = a_2 + jb_2. \qquad (1.6\text{ b})$$

The associative properties of algebra tell us we can group numbers however we see fit, e.g.,

$$g + f + h + j = (g+h) + (f+j), \tag{1.7}$$

so it makes sense to group the addition of complex numbers and to separate the j terms,

$$x_1 + x_2 = a_1 + jb_1 + a_2 + jb_2 = (a_1 + a_2) + j(b_1 + b_2). \tag{1.8}$$

Obviously the same approach can be used for subtraction. Now we look at the multiplication of two complex numbers:

$$x_1 x_2 = (a_1 + jb_1)(a_2 + jb_2) = ab + jb_1 a_2 + ja_1 b_2 + j^2 b_1 b_2. \tag{1.9}$$

Since $j = \sqrt{-1}$, we know $j^2 = -1$. Therefore, we can now regroup Eq. (1.9) in the following way,

$$x_1 x_2 = (a_1 + jb_1)(a_2 + jb_2) = ab - b_1 b_2 + j(b_1 a_2 + a_1 b_2), \tag{1.10}$$

i.e., another complex number. Division with complex numbers takes a little manipulation, but it still works.

$$\frac{x_1}{x_2} = \frac{(a_1 + jb_1)(a_2 - jb_2)}{(a_2 + jb_2)(a_2 - jb_2)} = \frac{a_1 a_2 + b_1 b_2 + j(b_1 a_2 - a_1 b_2)}{a_2^2 - b_2^2}. \tag{1.11}$$

The procedure of Eq. (1.11) exploits the fact that when you multiply a number by its complex conjugate, you get a real number.

1.1.2 Euler's Equation

So far, we are doing pretty well showing that complex numbers follow the same rules that we established for real numbers. What happens when we want evaluate an expression like Eq. (1.12)?

$$y = 10^{a+jb}. \tag{1.12}$$

We can write y as $y = 10^a 10^{jb}$. We know how to get 10^a when a is a real number, so now we focus on 10^{jb}. Until now, our approach has been to take the operations for real numbers and see if they hold for imaginary numbers containing j. We could use any base, but we use ten because it is the most familiar to us. We already know that

$$10^0 = 1,$$

so we start by looking at 10^ε when ε is a small number:

$$10^{0.01} = 1.0239299$$
$$10^{0.001} = 1.0023052$$
$$10^{0.0001} = 1.0002302$$

A pattern starts to emerge. It turns out that for small ε

$$10^\varepsilon \cong 1 + 2.3025\varepsilon. \tag{1.13}$$

We might use the following to check the accuracy. We know that

$$\left(10^{.0001}\right)^{10,000} = 10^1 = 10.$$

So using our approximation

$$\left(10^{.0001}\right)^{10000} = (1 + 0.00023025)^{10000} = 9.9965, \tag{1.14}$$

which indicates a fairly good approximation.

Previously, when we wanted to evaluate an operation using complex numbers we have used the same operations that apply to real numbers, keeping in mind Eq. (1.2) when we get multiples of j. We continue in this manner and start by evaluating 10^{js} when s is a small number. We will begin with $s = 0.001$, which gives

$$10^{j(0.001)} = 1 + j0.0023025. \tag{1.15}$$

To evaluate for $s = .002$, we will simply square the above term,

$$10^{j(0.002)} = \left(10^{j(0.001)}\right)^2 = (1 + j0.0023025)^2$$
$$= 1 + j0.0046 - (0.0023025)^2$$
$$= 1 - 0.00000529 + j0.0046$$

Rather than continue this process by hand, we can write a little

MATLAB program to determine 10^{js}:

```
for n=1:4000
s(n) = 0.001*n;
y(n) = 10^(j*s(n));
end
```

The results are shown in Fig. 1.3.

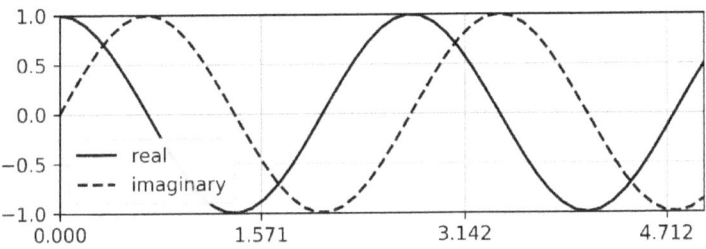

Figure 1.3. Graph of 10^{js}.

Notice that the pattern in Fig. 1.3 repeats itself. In particular

$$10^{i2.729} = 1.000 + i0.000.$$

That is interesting. It would probably be more interesting if this period corresponded to something physical, such as a circle. What if it repeated every 2π corresponding to the circumference of a circle? In other words, we want to find a such that,

$$10^{i2.729} = a^{i2\pi} = a^{i6.28318}.$$

If we take the base 10 logarithm of both sides

$$i2.729 = i6.28318 \cdot \log_{10} a$$

then,

$$\log_{10} a = \frac{2.729}{6.28318} = 0.43433.$$

The inverse logarithm give me the value of a,

$$a = 10^{0.43433} = 2.71828. \tag{1.16}$$

In fact, we will call this constant $e = 2.71828$. Now we redo Fig.1.3 to find e^{js} with the following MATLAB program:

```
e = 2.7182818;
for n=1:1000
s(n) = 0.001*n;
y(n) = e^(j*s(n));
end
```

The results are shown in Fig. 1.4.

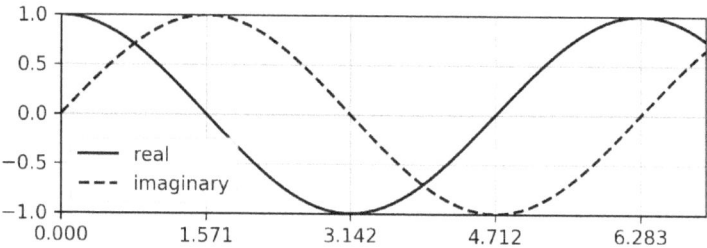

Figure 1.4. Graph of e^{js}

Notice that the real part goes from 1 to -1 and back to 1 in 2π radians, just like a cosine function; the imaginary part goes from zero to one to minus one and back to zero in 2π radians, just like a sine function. If that is the case, we might as well write

$$e^{j\theta} = \cos\theta + j\sin\theta, \tag{1.17}$$

where we have replaced s with θ. This is referred to as the *Euler's equation*, after the Swiss mathematician Leonhard Euler.

The Euler's equation is often called *the most beautiful equation*. You have probably studied analytic geometry, which was formulated by René Descartes, the French mathematician and philosopher. Descartes showed the link between algebra and geometry. Euler's formula is the link between algebra and trigonometry. You may or may not think it is the most beautiful equation, but as an electrical engineer, this equation will unquestionably prove to be the most important.

Consider a graph of Eq. (1. 17) as shown in Fig. 1. 5. Notice that the magnitude of $|e^{j\theta}| = 1$ so it plots a unit circle.

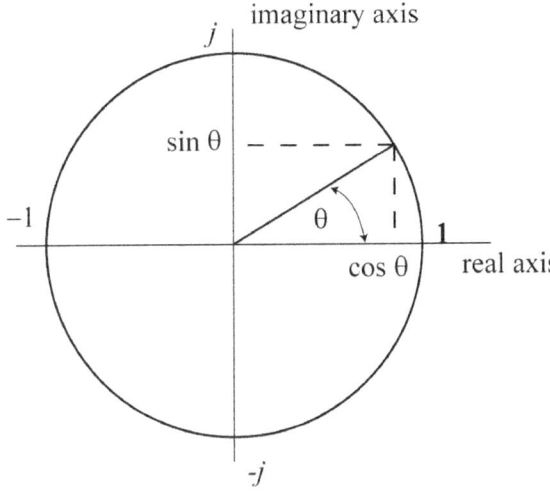

Figure 1.5. Graph of Eq. (1.1.17) in the complex plane.

We can expand Fig. 1.5. to include the entire complex plane. Figure 1.6 is a graph of the complex number, $z = 3 + j4$. We can describe that same point by specifying the magnitude and phase,

$$|z| = (3^2 + 4^2)^{1/2} = 5, \qquad (1.18\text{ a})$$

$$\angle z = \tan^{-1}\left(\frac{4}{3}\right) = 53°. \qquad (1.18\text{ b})$$

Therefore, we can say $z = 3 + j4 = 5e^{j53°}$. We refer to these as *polar coordinates*.

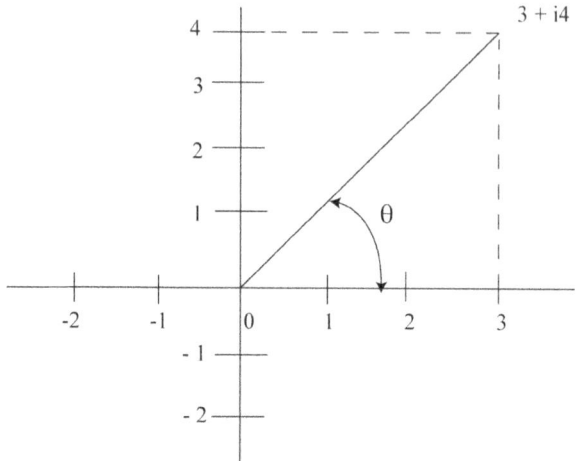

Figure 1.6. A graph of the complex number $z = 3 + j4 = 5e^{j53^\circ}$

Notice that if we think of j as having magnitude of one and an angle of 90 degrees,

$$j = 1\angle 90^\circ.$$

then,

$$j^2 = 1\angle 90^\circ \cdot 1\angle 90^\circ = 1\angle 180^\circ = -1.$$

We could have started with $-j = 1\angle -90^\circ$ and said

$$(-j)^2 = 1\angle -90^\circ \cdot 1\angle -90^\circ = 1\angle -180^\circ = -1.$$

What happens if we take the natural logarithm of a complex number in polar form?

$$\ln(z) = \ln(|z|e^{j\theta}) = \ln(|z|) + j\theta, \quad (1.19)$$

i.e. the real part is the magnitude, and the imaginary part is the phase. Remember: the θ in Eq. (1.19) is in radians.

Example 1.1. Solve for w.

$$e^w = 3 + j4.$$

Solution

$$\ln(e^w) = w = \ln(|w|e^{j\theta}) = \ln(3^2 + 4^2)^{1/2} + j\tan^{-1}\left(\frac{4}{3}\right) = 1.609 + j0.927$$

Check:
$$e^{(1.609+j0.927)} = e^{1.609}e^{j0.927} = 5(\cos(0.927) + j\sin(.927))$$
$$= 5(0.6 + j0.8) = 3 + j4$$

Equation (1.17) usually appears as one of a group of three equations. They are called the *Euler equations:*

$$e^{j\theta} = \cos\theta + j\sin\theta, \qquad (1.20\ \text{a})$$

$$\cos\theta = \frac{e^{j\theta} + e^{-j\theta}}{2}, \qquad (1.20\ \text{b})$$

$$\sin\theta = \frac{e^{j\theta} - e^{-j\theta}}{2j}. \qquad (1.20\ \text{c})$$

Can you prove Eq. (1.20 b & c) using just Eq. (1.20a)?

As an example of a typical use of Euler's equations, we might have a vector

$$A = 3.6e^{j34°}.$$

We can use Euler's equations to expand A

$$A = 3.6e^{j34°} = 3.6(\cos 34° + j\sin 34°)$$
$$= 3.6(0.829 + j0.574)$$
$$= 3 + j2$$

In using the polar coordinates, remember that angles can be in degrees or radians. This should not pose a problem, as long as you remember that $\pi = 180°$. For instance, if we have

$$X = 0.5e^{j3\pi/4},$$

but if the calculator uses degrees, make the following substitution:

$$X = 0.5e^{j3\pi/4} = 0.5e^{j3(180°)/4}$$
$$= 0.5e^{j135°} = -0.35 + j0.35$$

Besides the Euler equations above, the following trigonometric identities will prove to be very useful:

$$\sin(\alpha \pm 90) = \pm\cos\alpha, \qquad (1.21\text{ a})$$
$$\cos(\alpha \pm 90) = \mp\sin\alpha, \qquad (1.21\text{ b})$$
$$\sin(\alpha \pm \beta) = \sin\alpha\cos\beta \pm \cos\alpha\sin\beta, \qquad (1.21\text{ c})$$
$$\cos(\alpha \pm \beta) = \cos\alpha\cos\beta \mp \sin\alpha\sin\beta. \qquad (1.21\text{ c})$$

1.1.3 Phasor Notation

Suppose we are talking about a group of sinusoidal functions at one frequency:

$$y_1(t) = 2\cos(\omega t + 10°), \qquad (1.22\text{ a})$$
$$y_2(t) = 4\cos(\omega t - 40°), \qquad (1.22\text{ b})$$
$$y_3(t) = \sin(\omega t + 20°). \qquad (1.22\text{ c})$$

Notice that we can write

$$y_1(t) = \text{Re}\left\{e^{j(\omega t + 10°)}\right\},$$
$$y_2(t) = \text{Re}\left\{e^{j(\omega t - 40°)}\right\},$$

$$y_3(t) = \sin(\omega t + 20°)$$
$$= -\cos(\omega t + 20° - 90°) = \text{Re}\left\{1e^{j(\omega t - 70°)}\right\}$$

Each of these has the term $e^{j\omega t}$. In each case, we will take the real part to get back to the cosine functions. We could define a new set of complex numbers, called *phasors*, that specify the amplitudes and phases of these functions:

$$Y_1 = 2e^{j10°} \qquad (1.23\text{ a})$$
$$Y_2 = 4e^{-j40°} \qquad (1.23\text{ b})$$
$$Y_3 = -1e^{-j70°} = 1e^{j110°}. \qquad (1.23\text{ c})$$

Each of the functions of Eq. (1.23) has been reduced to one complex number, the phasor. Notice that a phasor is always defined in the context of one frequency and either a sine or cosine function. We can conduct operations on the phasors that are much simpler than using the same operations on the functions.

Example 1.2. Calculate

$$y_4(t) = 2\cos(\omega t + 10°) - \sin(\omega t + 20°).$$

Solution

$$-\sin(\omega t + 20°) = -\cos(\omega t + 20° - 90°) = -\cos(\omega t - 70°).$$

Using phasors we can calculate,

$$Y_4 = Y_1 - Y_3 = 2e^{j10°} - 1e^{-j70°} = 2.075e^{j38°}.$$

This is a simple operation with MATLAB or Python or any program that has complex variables. We go back to the time domain with,

$$y_4(t) = 2.075\cos(\omega t + 38°).$$

Notice that we did not even know the frequency ω to calculate $y_4(t)$.

1.1.4 Using Phasors to Solve Differential Equations with Sinusoidal Inputs

Suppose we are given a differential equation in the form

$$\frac{dy(t)}{dt} + 1.2y(t) = \cos(3t). \qquad (1.24)$$

Here is a very simple principle that will prove very helpful is solving differential equations with sines or cosines as the forcing functions:

The solution to a differential equation with a sinusoidal forcing function must be a sinusoid at the same frequency.

In other words, since the forcing function for the above equation is $f(t) = \cos(3t)$, the solution must be in the form $y(t) = A\cos(3t + \theta)$. The amplitude A can change and the phase can be shifted, but the frequency $\omega = 3$ cannot change.

One way we can solve this is to say that the solution must have the form

$$y(t) = A\cos(3t) + B\sin(3t).$$

The first derivative is

$$\frac{dy(t)}{dt} = -3A\sin(3t) + 3B\cos(3t).$$

Now the equation is

$$[-3A\sin(3t) + 3B\cos(3t)] + 1.2[A\cos(3t) + B\sin(3t)] = \cos(3t).$$

For the equation to be true, all the cosine and sine terms have to balance:

$$3B\cos(3t) + 1.2A\cos(3t) = \cos(3t)$$
$$-3A\sin(3t) + 1.2B\sin(3t) = 0$$

or

$$3B + 1.2A = 1,$$
$$-3A + 1.2B = 0.$$

The second equation gives

$$A = \frac{1.2}{3}B = 0.4B,$$

so the first becomes
$$3B + 1.2(0.4B) = 1,$$
$$3B + 0.48B = 1, \quad B = 0.287, \quad A = 0.115.$$

The solution is
$$y(t) = 0.115\cos(3t) + 0.287\sin(3t)$$
$$= 0.309\cos(3t - 68°).$$

That was straight-forward, but very laborious. Suppose we go back and write Eq. (1.24) as
$$\frac{dy(t)}{dt} + 1.2y(t) = \text{Re}\{e^{j3t}\}.$$

Let us instead solve the following equation:
$$\frac{dy(t)}{dt} + 1.2y(t) = e^{j3t},$$

Then in the end, we will just take the real part of the solution to get back to cosines. We know the answer $y(t)$ must be a sinusoid with radian frequency 3, so we can write
$$y(t) = Ye^{j3t}.$$

When we put this in the differential equation, we get
$$\frac{d}{dt}(Ye^{j3t}) + 1.2(Ye^{j3t}) = \{e^{j3t}\},$$

or
$$3jYe^{j3t} + 1.2Ye^{j3t} = e^{j3t}.$$

In fact, since the e^{j3t} is common to every term, we can divide it out and rewrite the equation as
$$3jY + 1.2Y = 1.$$

Solving for the phasor Y,
$$Y = \frac{1}{1.2 + 3j} = \frac{1}{3.23\angle 68°} = 0.309e^{-j68°}.$$

The answer is

$$\text{Re}[y(t)] = \text{Re}[Ye^{j3t}]$$
$$= \text{Re}[0.309e^{-j68°}e^{j3t}] = 0.309\cos(3t-68°).$$

Suppose the input had been
$$\cos(3t+10°) = \text{Re}(e^{j3t+j10°}).$$

We could go back and redo everything, but that is not really necessary. The phasor equation is
$$3jYe^{j3t} + 1.2Ye^{j3t} = e^{j(3t+10°)},$$

and so
$$Y = \frac{e^{j10°}}{1.2+3j} = \frac{e^{j10°}}{3.23\angle 68°} = 0.309e^{-j58°}.$$

What if the input had been
$$x(t) = \sin(3t)?$$

Then we use
$$\text{Im}[e^{j3t}] = \sin(3t);$$

but the mathematics will all be the same.

Now suppose we have an input in the form
$$\frac{dy(t)}{dt} + 1.2y(t) = \cos(3t+10) + 5\sin(3t-30°).$$

Linearity tells us we can just solve two separate problems and then put them together

$$y(t) = 0.309\cos(3t+10°-68°) + 5(.309)\sin(3t-30°-68°)$$
$$= 0.309\cos(3t-58°) + 1.545\sin(3t-98°).$$

However, if the input is one frequency, we only want to solve for one input. The following two formulas, derived from Eq. (1.21 a) and (1.21b) are often very helpful:

$$\sin(\alpha) = -\cos(\alpha+90),$$
$$\cos(\alpha) = \sin(\alpha+90).$$

We could convert either, but we would rather work with cosines, so

$$5\sin(3t-30°) = -5\cos(3t-30°+90°) = -5\cos(3t-90°),$$

and the input is

$$x(t) = \cos(3t+10) - 5\cos(3t-60°).$$

Writing this as a phasor, we get

$$X = 1e^{j10°} - 5e^{-j60°} = (.89 + j0.17) - (2.5 - j4.33)$$

$$= -1.61 + j4.5 = 4.77e^{j110°}.$$

Therefore, the output will be

$$Y = 0.309e^{j(-58°)} 4.47e^{j110°}$$

$$= 1.38e^{j52°},$$

$$y(t) = 1.38\cos(3t + 52°).$$

Example 1.3. Solve for y(t):

$$2\frac{dy(t)}{dt} + 4y(t) = 10\sin(2t).$$

Solution

First, divide through by 2 so the leading coefficient is 1:

$$\frac{dy(t)}{dt} + 2y(t) = 5\sin(2t).$$

In phasor notation, the original equation becomes

$$j2Y + 2Y = 5.$$

Now just solve for Y:

$$Y = \frac{5}{2+j2} = \frac{5}{1.41\angle 45°} = 3.55e^{-j45°}.$$

This results in,

$$y(t) = 3.55\sin(2t - 45°).$$

Example 1.4. Solve the following differential equation.

$$\frac{dy(t)}{dt} + 2y(t) = 3\sin(2t - 63°).$$

Solution

Clearly, this is the same problem with a different magnitude and phase of the forcing function. Linearity dictates that the answer is the same as above, with the amplitude changed by

$$\frac{3}{5} = 0.6,$$

and phase changed by $-63°$, or

$$y(t) = (0.6)3.55\sin(2t - 45° - 63°)$$
$$= 2.13\sin(2t - 108°).$$

Check:

$$Y = \frac{3\angle -63°}{2+j2} = \frac{3\angle -63°}{1.41\angle 45°} = 2.13e^{-j108°}$$

Example 1.5. Solve for $y(t)$. The answer should be only sines.

$$\frac{dy(t)}{dt} + 3y(t) = \cos(3t) + \sin(3t)$$

Solution

$$\cos(3t) + \sin(3t) = \sin(3t + 90°) + \sin(3t)$$
$$= \text{Im}\left[e^{j3t}e^{j90°} + e^{j3t}\right] = \text{Im}\left[(1+j)e^{j3t}\right]$$

We assume the solution is in the form

$$y(t) = \text{Im}\left[Ye^{j3t}\right],$$

so the original equation becomes

$$j3Ye^{j3t} + 3Ye^{j3t} = (1+j)e^{j3t}$$

$$j3Y + 3Y = (1+j)$$

$$Y = \frac{1+j}{3(1+j)} = \frac{1}{3}.$$

Therefore, the answer is

$$y(t) = \frac{1}{3}\sin(3t)$$

1.5 Summary of Phasors

If you still have trouble understanding phasors to solve differential equations in the form

$$\frac{d^2y(t)}{dt^2} + 3\frac{dy(t)}{dt} + 2y(t) = 3\sin(2t + 45°), \qquad (1.25)$$

then follow this procedure:

1. Replace the input with an exponential of the same frequency, .i.e., $f(t) = e^{j2t}$.

2. Assume the solution is in the form $y(t) = Ye^{j2t}$; Y is a complex number.

3. Solve for Y using the $y(t)$ from step 2 and the input $f(t)$ from step 1; convert Y to polar coordinates

$$Y = |Y|e^{j\theta_Y}.$$

4. The solution will be the original input attenuated by $|Y|$ as well as the magnitude of the forcing function, and phase shifted by θ_Y as well as the phase of the forcing function.

Example 1.6. Solve equation (1.25) using phasors.

Solution

Step 3 above is

$$\frac{d^2}{dt^2}(Ye^{j2t}) + 3\frac{d}{dt}(Ye^{j2t}) + 2(Ye^{j2t}) = e^{j2t}.$$

Taking the derivatives gives:

$$(j2)^2 Y e^{j2t} + 3(j2) Y e^{j2t} + 2Y e^{j2t} = e^{j2t}.$$

We can divide out a factor e^{j2t} leaving:
$$(j2)^2 Y + 3(j2)Y + 2Y = 1.$$

Now we can solve for Y:
$$-4Y + j6Y + 2Y = 1,$$
$$Y = \frac{1}{-2+j6} = \frac{1}{\sqrt{40} \angle 124°} = 0.158 e^{-j124°}.$$

So the answer is
$$y(t) = 3(0.158) \sin(2t + 45° - 124°)$$
$$= 0.474 \sin(2t - 79°)$$

If the input is a cosine instead of a sine, we use exactly the same procedure, except that the answer will also be a cosine. Notice that it does not matter what the order of the differential equation is. Y will always be one complex number.

The following procedure is an even simpler way to use phasors.

For the equation
$$\frac{d^2 y(t)}{dt^2} + 3 \frac{dy(t)}{dt} + 2y(t) = 3\sin(2t + 45°)$$

1. Create a new algebraic equation:
$$(j\omega)^2 Y + 3(j\omega)Y + 2Y = 1$$

2. Solve for the complex variable Y
$$Y = \left. \frac{1}{(j\omega)^2 + 3(j\omega) + 2} \right|_{\omega=2} = \frac{1}{-4+j6+2} = \frac{1}{\sqrt{40}\angle 124°} = 0.158 e^{-j124°}$$

3. The answer is
$$y(t) = 3|Y|\sin(2t + 45° + \angle Y)$$
$$= 3(0.158)\sin(2t + 45° - 124°)$$
$$= 0.474 \sin(2t - 79°)$$

1.2 Signals

1.2.1 Introduction to Signals

The function *f(t)* in Fig. 1.7 is a *signal*. It could represent anything: a current, a voltage, the water flow from a pipe, or the mean temperature at the North Pole. *f(t)* is continuous, i.e., it is a mathematical function of an independent variable *t*. Note that continuous also means that it is uniquely defined in *t*, except for maybe a finite number of points, as shown in Fig. 1.8.

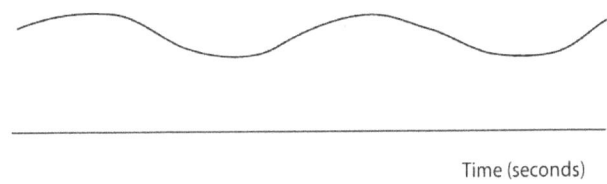

Time (seconds)

Figure 1.7. A continuous signal

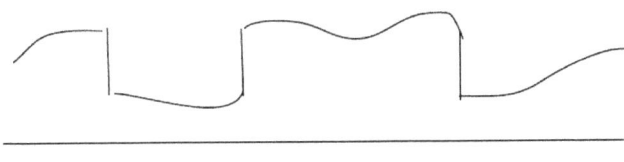

Time (seconds)

Figure 1.8. A continuous signal with a finite number of discontinuities.

The function $g(t) = \sqrt{t}$ does not qualify as a signal because the square root has two values. Signals that represent real quantities in the time domain are real. We deal with complex numbers only after making a transformation to the frequency domain, e.g., the Laplace, Fourier, or Z domain.

1.2.2 Special Signals

We are all familiar with signals like $\sin(5t)$, e^{-3t}, etc. These are *ordinary* signals. These are signals that we can generate with a function generator in the lab. But another group of signals are called *distributions,* or *singular functions.* These are often ideal functions that we may or may not be able to generate.

1. Step Function

The first of these new functions is called the step function. It is one that we have used already:
$$u(t) = 0 \quad t < 0,$$
$$ = 1 \quad t \geq 0. \tag{1.26}$$

This is a signal that we cannot truly generate, because we cannot turn anything on infinitely fast. We can come very close. The step function is important in its role of defining *causal* functions. Causal functions are those that are zero before time t = 0. This is often an important distinction. For instance, suppose we were given the following differential equations

$$\frac{dy(t)}{dt} + 0.5 y(t) = \sin(3t) \tag{1.27 a}$$

$$\frac{dy(t)}{dt} + 0.5 y(t) = \sin(3t)\, u(t) \tag{1.27 b}$$

Equation (1.27 a) would be solved with phasors, whereas Eq. (1.27 b) would be best solved with Laplace transforms.

There is an alternate to the step function, called the *Heavyside* function, which is sometimes used
$$u_h(t) = 0 \quad t < 0,$$
$$ = 0.5 \quad t = 0, \tag{1.28}$$
$$ = 1 \quad t > 0.$$

This has certain mathematically desirable properties that are used when dealing with things like inverse Fourier transforms. However, the vast majority of the time, $u_h(t)$ may be considered interchangeable with $u(t)$.

2. Sigmun function.

This is closely related to the step function.
$$\operatorname{sgn}(t) = -1 \quad t < 0,$$
$$\phantom{\operatorname{sgn}(t)} = 1 \quad t > 0. \tag{1.29}$$

In fact, the Heavyside function is often defined as
$$u_h(t) = \frac{1}{2} + \frac{1}{2}\operatorname{sgn}(t). \tag{1.30}$$

3. Ramp function

The ramp function is given by
$$r(t) = 0 \quad t < 0$$
$$= t \quad t \geq 0 \quad (1.31\ a)$$

The ramp function may be thought of as the integral of the step function
$$r(t) = \int_{-\infty}^{t} u(\tau)d\tau \quad (1.31\ b)$$

{The tau τ in the integral illustrates the use of a dummy variable. Like t, it is a function of time, but it is not the "real t" that appears in $r(t)$. It is just a parameter being used to do the integration.}

4. Rectangular pulse $p_\tau(t)$

$$p_\tau(t) = 1 \quad -\tau/2 \leq t \leq \tau/2$$
$$= 0 \quad elsewhere \quad (1.32)$$

5. Triangular pulse $\Delta_\tau(t)$

$$\Delta_\tau(t) = 0 \quad t \leq -\tau/2,$$
$$= 1 + 2t/\tau \quad -\tau/2 \leq t \leq 0,$$
$$= 1 - 2t/\tau \quad 0 \leq t \leq \tau/2, \quad (1.33)$$
$$= 0 \quad \tau/2 \leq t.$$

6. Sinc function sinc(t)

$$\text{sinc}(t) = \frac{\sin(t)}{t}. \quad (1.34)$$

{Be aware that the sinc function is also often defined as
$\text{sinc}(t) = \dfrac{\sin(\pi t)}{\pi t}$. However, we will use Eq. (1.30) in this book.}

Example 1.7. Represent the following signals with mathematical expressions.

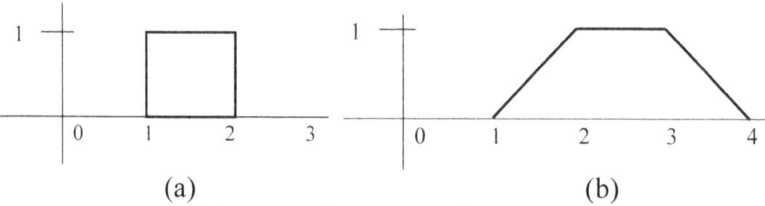

(a)   (b)

Figure 1.19 Some continuous signals

Solution

(a) We can write this as $x_a(t) = u(t-1) - u(t-2)$, or as

$$x_a(t) = p_1(t-1.5)$$

(b) $x_b(t) = r(t-1) - r(t-2) - r(t-3) + r(t-4)$

---

7. The impulse delta function

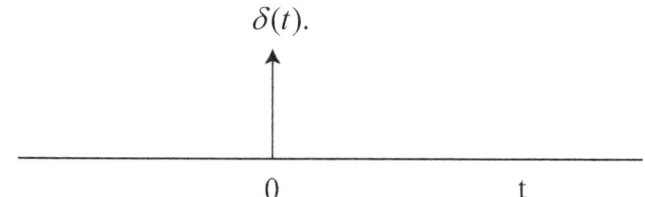

Figure 1.10. The delta function $\delta(t)$.

The delta function has a value at one point only, as shown in Fig. 1.10. The amplitude of the delta function is defined as follows:

$$\int_{-\infty}^{\infty} \delta(t)\, dt = \int_{0^-}^{0^+} \delta(t)\, dt = 1. \qquad (1.35)$$

We could define it from the pulse function in a limiting sense,

$$\delta(t) = \lim_{\tau \to 0} \left\{ \frac{1}{\tau} p_\tau(t) \right\}.$$

In other words, we can think of it as a narrow rectangle with an area of one. One of the most important properties involving the delta function is the following:

$$\int_{-\infty}^{\infty} f(t)\delta(t-t_0)\,dt = f(t_0). \tag{1.35}$$

Since $\delta(t-t_0)$ is only defined at one point, t_0, the entire integral is only evaluated at one point. This is often called the *sifting property*.

Example 1.8. Evaluate $f(t)$:

$$f(t) = \int_{-\infty}^{t} \delta(\tau)d\tau.$$

Solution

Think about integrating the delta function in Fig. 1.2.2. As we start from minus infinity and move towards zero, obviously the integral is zero. Then at t = 0, the integral jumps to one as a result of Eq. (1.2.10). In other words, this is the step function,

$$u(t) = \int_{-\infty}^{t} \delta(\tau)d\tau. \tag{1.36}$$

It is important to notice that when the dummy variable τ is used in the integral and the upper limit is time, t, the result of the integral is another function of time, not just a number. Note also that as a result of Eq. (1.33), we know that

$$\delta(t) = \frac{du(t)}{dt}.$$

There is a similar mathematical definition that is used:

$$\int_{-\infty}^{t_0} f(t)\delta(t-t_0)\,dt = \frac{1}{2}f(t_0).$$

This is a classic case of making a definition consistent with other things, since

$$\int_{-\infty}^{t_0} f(t)\delta(t-t_0)\,dt + \int_{t_0}^{\infty} f(t)\delta(t-t_0)\,dt = \int_{-\infty}^{\infty} f(t)\delta(t-t_0)\,dt$$

Time Scaling Property

$$\int_{-\infty}^{\infty} f(t)\delta(at - t_0) \, dt = \frac{1}{|a|} f\left(\frac{t_0}{a}\right)$$

$$\Rightarrow \delta(at - t_0) = \frac{1}{|a|} \delta\left(t - \frac{t_0}{a}\right)$$

(1.37)

Proof First suppose a>0, and introduce a change of variables

$$\sigma = at - t_0 \Rightarrow t = \frac{\sigma + t_0}{a}, \quad dt = \frac{d\sigma}{a}$$

$$\int_{-\infty}^{\infty} f(t)\delta(at - t_0) \, dt = \int_{\sigma=-\infty}^{\sigma=\infty} f\left(\frac{\sigma + t_0}{a}\right) \delta(\sigma) \frac{d\sigma}{a} = \frac{1}{a} f\left(\frac{t_0}{a}\right)$$

For a<0, the integral above is the same except for the limits of integration

$$\int_{-\infty}^{\infty} f(t)\delta(at - t_0) \, dt = \int_{\sigma=\infty}^{\sigma=-\infty} f\left(\frac{\sigma + t_0}{a}\right) \delta(\sigma) \frac{d\sigma}{a} = -\frac{1}{a} f\left(\frac{t_0}{a}\right)$$

which proves Eq. (1.2.13).

Derivative Property.

We define a new function, the derivative of the impulse,

$$\delta^{(1)}(t) = \frac{d}{dt}\delta(t).$$

Now we have the following formula:

$$\int_{-\infty}^{\infty} f(t)\delta^{(1)}(t - t_0) \, dt = -f^{(1)}(t_0)$$

(1.38)

Proof. Integration by parts yields

$$\int_{-\infty}^{\infty} f(t)\delta^{(1)}(t - t_0) \, dt = f(t)\delta(t - t_0)\Big|_{-\infty}^{\infty}$$

$$- \int_{-\infty}^{\infty} f^{(1)}(t)\delta(t - t_0) \, dt = -f^{(1)}(t_0)$$

The function $f(t)\delta(t - t_0)$ is zero at plus or minus infinity.

Similar proofs lead to similar formulas for other values of n:

$$\int_{-\infty}^{\infty} f(t)\delta^{(n)}(t - t_0) \, dt = \left(-1^n\right) f^{(n)}(t_0).$$

(1.39)

Example 1.9. Look at the following integrals and be sure you understand the given results.

$$\int_{-\infty}^{\infty} \delta(t)u(t-2)\, dt = 0,$$

$$\int_{-\infty}^{\infty} \delta(t)u(t+2)\, dt = 1,$$

$$\int_{-\infty}^{\infty} \delta^{(1)}(t)\, dt = \delta(t).$$

This result is purely from the definition of $\delta^{(1)}(t)$.

$$\int_{-\infty}^{3} e^{-2t}\delta(t-3)\, dt = \frac{1}{2}e^{-6},$$

$$\int_{-\infty}^{\infty} e^{-2t}\delta^{(1)}(t-3)\, dt = -\frac{d}{dt}\left(e^{-2t}\right)_{t=3} = 2e^{-6},$$

$$\int_{-\infty}^{\infty} e^{-2t}\delta^{(4)}(t-3)\, dt = (-1)^{4}\frac{d^{4}}{dt^{4}}\left(e^{-2t}\right)_{t=3}$$

$$= (-2)^{4} e^{-2t}\Big|_{t=3} = 16e^{-6}.$$

Example 1.10. Simplify the following expression as much as possible:

$$\cos(0.5\pi t)\delta(t+1) + \int_{-\infty}^{\infty} e^{-5t}\left[\delta(3t-2) + \delta^{(1)}(t-2)\right] dt$$

$$+ \int_{-5}^{6} 3\delta(t+3)\, dt$$

Solution

$$\cos(-0.5\pi)\delta(t+1) + \frac{1}{3}e^{-5(2/3)} + (-1)(-5)e^{-5(2)} + 3$$

Exercises:

1.1 Complex Numbers

1.1.1. Using these complex numbers,

$$A = 3 + j4$$

$B = 3 - j5$

$C = 0.5 / \_45^0$

$D = 2.0e^{-j\pi/6}$

Find:

1. $C + D$
2. $C + D^*$
3. $A \cdot B$
4. B / A
5. $A + \dfrac{C}{D^*}$
6. $A\left(C + \dfrac{D}{B^*}\right)$

Show each step of the calculation. Any addition or subtraction must appear in rectangular coordinates and any multiplication or division must appear in polar coordinates.
{Note: '*' means the complex conjugate, i.e., if $x = 4 + j2$, then $x^* = 4 - j2$}

1.1.2. Compute e^z (in the form $u + jv$) for

(a) $z = -j$ (b) $z = -2 - j3\pi$ (c) $z = e + j5\pi$ (d) $z = -1 - j7\pi/4$

1.1.3. Express the following as just one sine or cosine term by first converting the time-domain functions to phasors.

a. $x_1(t) = 3\sin(2t + 45^\circ) + 2\sin(2t - 35^\circ)$

b. $x_2(t) = 3\cos(2t + 45^\circ) + 2\sin(2t - 35^\circ)$

1.1.4. Solve for $y(t)$ for the given inputs $f(t)$

$$\dfrac{d^2 y(t)}{dt^2} + 2\dfrac{dy(t)}{dt} + 4y(t) = f(t)$$

a. $f(t) = 3\cos(5t - 10^\circ)$

b. $f(t) = 5\sin(5t + 45°)$

c. $f(t) = 2\cos(2t + 178°)$

1.1.5. Prove the following using Euler's equations:
$$\sin^2(\theta) = \frac{1}{2} - \frac{1}{2}\cos(2\theta)$$

1.1.6. A system is described by the following differential equation

$$\frac{d^3}{dt^3}y(t) + 2\frac{d^2}{dt^2}y(t) + 9\frac{d}{dt}y(t) + 2y(t) = f(t)$$

Solve for $y(t)$ if $f(t) = 3\sin(t) + 2\cos(3t)$. (Your answer should consist of one sine term and one cosine term).

1.1.7 Write the following equation as just a cosine with a phase term:
$$y(t) = A\cos(\omega t) + B\sin(\omega t).$$

1.2 Signals

1.2.1 Evaluate the following. Write the answer in the most concise form possible.

a. $\int_{-3}^{10} \left[\sin\left(\frac{2\pi t}{5}\right) + \cos\left(\frac{2\pi t}{5}\right)\right]\delta\left(t - \frac{5}{2}\right) dt$

b. $\left[\sin\left(\frac{2\pi t}{5}\right) + \cos\left(\frac{2\pi t}{5}\right)\right]\delta\left(t - \frac{5}{2}\right)$

1.2.2. Write each of the following in the simplest form:

a. $\int_0^t e^{-5(t-\tau)}\delta(\tau)d\tau$

b. $\left(\cos(10^{-4}t) + \sin(10^2 t)\right)\delta(t)$

c. $\int_{-\infty}^{4} e^{-2t} r(t) \delta(t-2) dt$

d. $\int_{-\infty}^{4} [\delta(t+3) + \delta(t-3)] e^{-t} dt$

e. $t \cdot \delta(t-6)$

f. $\int_{-\infty}^{3} \delta^{(1)}(t-2) t^2 dt$

1.2.3. Describe the following functions mathematically.
a).

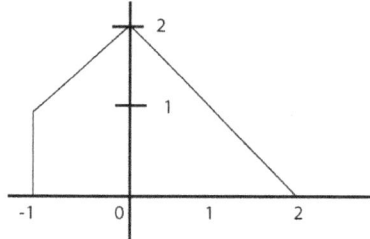

b).

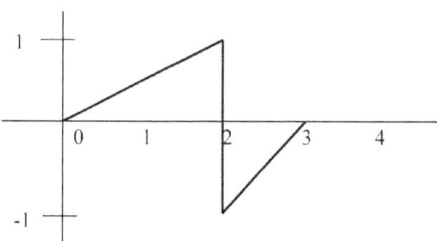

1.2.4. Sketch the following function. Label the axes.

$$f(t) = r(t) - u(t-1) - r(t-2) - r(t-3)$$

Chapter 2. Systems and Convolution

2.1 Linearity and Time Invariance

In this book, we will deal with continuous and discrete time dynamic systems with the following two important characteristics: the systems are *linear,* and are *time-invariant.* These two properties, *linearity* and *time-invariance*, are of enormous importance. Linearity basically means if you scale the input to a system by a certain amount, the output is scaled by the same amount. Likewise, if the input is the sum of two signals, then the output is the sum of the results from the individual inputs. Time invariance means that if the input is delayed by a certain time, then the output will be identical but delayed by the same time. The meanings are illustrated in Fig. 2.1

In engineering and science, we describe physical systems with differential equations. *Linear* means the differential equation has constant coefficients. *Time invariant* means that the coefficients do not change in time. As an example, Eq (2.1 a) is linear and time invariant. Equation (2. 1b) is not linear, because of the *3y* coefficient in front of the second term. Equation (2.1 c) is not time invariant because of the *3t* term in front of the second term.

$$\frac{d^2 y(t)}{dt^2} + 3\frac{dy(t)}{dt} + 2y(t) = f(t) \quad\quad (2.1\ \text{a})$$

$$\frac{d^2 y(t)}{dt^2} + 3y(t)\frac{dy(t)}{dt} + 2y(t) = f(t) \quad\quad (2.1\ \text{b})$$

$$\frac{d^2 y(t)}{dt^2} + 3t\frac{dy(t)}{dt} + 2y(t) = f(t) \quad\quad (2.1\ \text{c})$$

If

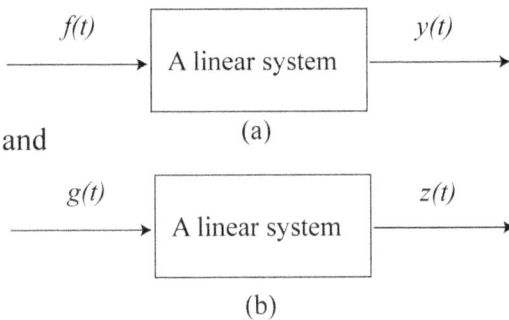

and

then *linearity* says:

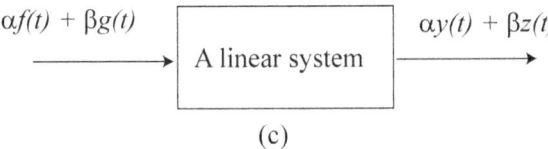

and *time invariance* says:

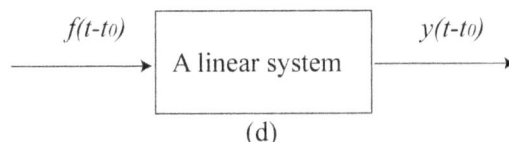

Figure 2.1. Illustration of linearity and time invariance

2.2 System Analysis

A *system* can be anything. It can be a circuit, a factory, an ecosystem, etc. Physically, systems are described using differential equations. We use the physical properties of the components of the system, along with the laws governing their interaction, to put together differential equations. For instance, look at the *RC* circuit in Fig. 2.2.

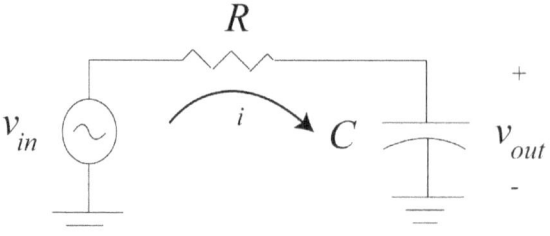

Figure 2.2. An RC circuit

The physical principles are the relationships between the voltage and the current for the individual components:

$$v_R = iR,$$

$$i = C \frac{dv_{out}}{dt}.$$

Kirchov's voltage law tells us that the sum of the voltages around the circuit must be zero, so

$$v_{in} = RC \frac{dv_{out}}{dt} + v_{out}.$$

We will find it more convenient to write it as follows,

$$\frac{dv_{out}}{dt} + \frac{1}{RC} v_{out} = \frac{1}{RC} v_{in}. \qquad (2.2)$$

Now we have a differential equation that describes this "system," a simple electric circuit with input $v_{in}(t)$. In actuality, there are two sources of energy that can contribute to v_{out}, one is the forcing function, $v_{in}(t)$, and the other is the initial condition. In this case there can be a charge on the capacitor. By linearity, we can analyze a circuit by finding the response to the initial condition, then finding the response to the forcing function, and adding them together to get the total response. We will start with the response to the initial condition.

2.2.1 The Response to the Initial Condition

If we are interested only in the response due to the initial condition, we can use the model in Fig. 2.3. Before the switch is closed, let us assume there is a voltage on the capacitor that we will write as $v_{out}(0^-)$. The homogeneous response of the first order differential equation in Eq. (2.2) is

$$v_{out}(t) = Be^{-t/RC} u(t). \tag{2.3}$$

The fact that we know the voltage at time $t=0$ means we can solve for A

$$v_{out}(0) = Be^{-0/RC} u(t) = B = v_{out}(0^-)$$

So we say that the *zero-input* solution is

$$v_{zi}(t) = v_{out}(0^-) e^{-t/RC} u(t).$$

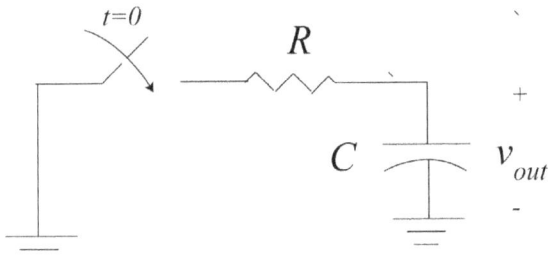

Figure 2.3. An RC circuit with an initial condition

2.2.2. The Impulse Response and Convolution

<u>The Impulse Response</u>

In Eq. (2.2), we want to solve $v_{out}(t)$ for any $v_{in}(t)$. Why not just put something in, and see what comes out? How about a delta function, $\delta(t)$? So we will try to solve,

$$\frac{dv_{out}(t)}{dt} + \frac{1}{RC} v_{out}(t) = \delta(t). \tag{2.4}$$

The first thing to do is determine what kind of response we are likely to get. To do this, we look at the characteristic equation,

$$s + \frac{1}{RC} = 0.$$

We got this by ignoring the forcing function and putting in an *s* for the derivative. The solution gives

$$s = -\frac{1}{RC}.$$

So we assume the solution has the form

$$v_{out}(t) = Ae^{-t/RC}u(t)$$

Note the *u(t)*, which means our function is causal. We expect that. Nothing is going to come out until we put something in. First, look at the derivative:

$$\frac{d}{dt}v_{out}(t) = A\left(\frac{-1}{RC}\right)e^{-t/RC}u(t) + Ae^{-t/RC}\delta(t).$$

Substituting this expression into Eq. (2.2.4) gives us

$$A\left(\frac{-1}{RC}\right)e^{-t/RC}u(t) + Ae^{-t/RC}\delta(t) + \frac{1}{RC}Ae^{-t/RC} = \delta(t). \quad (2.5)$$

Equation (2.5) has two different kinds of functions in it. For the equality to hold, the following two equations must hold:

The $\delta(t)$ terms give

$$A\delta(t) = \delta(t),$$

and the *u(t)* terms give

$$A\left(\frac{-1}{RC}\right)e^{-t/RC} + \frac{1}{RC}Ae^{-t/RC} = 0.$$

The only solution to both these equations is $A = 1$. This solution of $v_{out}(t)$ for the particular input of $\delta(t)$ is so important that it has its own name: the *impulse response h(t)*,

$$h(t) = e^{-t/RC}u(t). \quad (2.6)$$

Once we know the impulse response to a system, we can find the response to any input by the *convolution* of the input function with the impulse response.

The Convolution Integral

The following is an attempt to explain the origin of the convolution integral. Suppose we are interested in the area below $x(\tau)$ up to a time t,

$$I(t) = \int_0^t x(\tau) d\tau. \tag{2.7}$$

We might approximate this as is shown in Fig. 2.4.

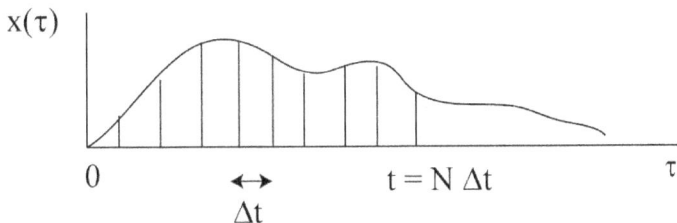

Figure 2.4. The area under the curve is approximated as the sum of areas as shown.

We could express Fig. 2.4 mathematically as follows,

$$I(t) \cong \sum_{n=0}^{N} x(n \cdot \Delta t) \Delta t. \tag{2.8}$$

Then as the Δt gets smaller, we can write

$$I(t) = \lim_{\Delta t \to 0} \sum_{n=0}^{N} x(n \cdot \Delta t) \Delta t = \int_0^t x(\tau) d\tau. \tag{2.9}$$

We have arrived at Eq. (2.9) by making the substitutions

$$\tau = \Delta t \cdot n, \tag{2.10 a}$$

and

$$d\tau = \lim_{\Delta t \to 0} \Delta t, \tag{2.10 b}$$

and then replacing the summation with an integral. Equation (2.9) is usually accepted as the definition of the integral. Note that $x(t)$ is a function of t, because the area under the curve varies with time.

Recall that in chapter 1 we said we could define the delta function as a pulse function in a limiting sense,

$$\delta(t) = \lim_{\tau \to 0} \left\{ \frac{1}{\tau} p_\tau(t) \right\}.$$

Therefore we could take any one of the little subsections in Fig. 2.4 and write it as

$$x_n = x(n \cdot \Delta t) \cdot \Delta t \, \delta(t - n \cdot \Delta t). \tag{2.11}$$

Let us now assume that we have a system with an impulse response illustrated in Fig. 2.5.

$$h(t) = e^{-\alpha t} u(t). \tag{2.12}$$

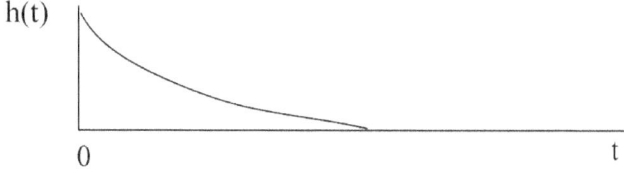

Figure 2.5. The impulse response of Eq. (2.12)

We know that $h(t)$ is the response to an impulse at time $t = 0$. If there is an impulse at time t_0, the impulse response is

$$h(t) = e^{-\alpha(t-t_0)} u(t - t_0), \tag{2.13}$$

as illustrated in Fig. 2.6.

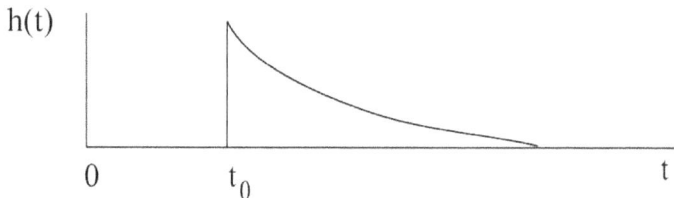

Figure 2.6. The impulse response delayed by time t_0.

If the function $x(\tau)$ of Fig. 2.4 is the input to the system with impulse response $h(t)$, then by virtue of linearity and time-invariance the output will be the sum of the all the impulse responses to all the individual pulses of Fig. 2.4,

$$y(t = N \cdot \Delta t) = \sum_{n=0}^{N} x(n \cdot \Delta t) \cdot \Delta t \left(e^{-\alpha(t - n \cdot \Delta t)} u(t - n \cdot \Delta t) \right). \quad (2.14)$$

In the limit as Δt gets smaller, we can write

$$y(t) = \lim_{\Delta t \to 0} y(t = N \cdot \Delta t) = \int_{0}^{t} x(\tau) e^{-\alpha(t - \tau)} d\tau, \quad (2.15)$$

which can be expressed as

$$y(t) = \int_{0}^{t} x(\tau) h(t - \tau) d\tau. \quad (2.16)$$

This is the convolution integral.

The graph of the function $x(\tau)$ is given in Fig. 2.5. The graph of $h(t - \tau)$ is shown in Fig. 2.7,

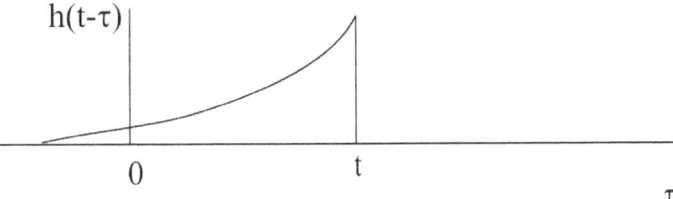

Figure 2.7. The graph of $h(t - \tau)$.

The graph of the term in the integrand in Eq. (2.16) is given in Fig. 2.8. Figure 2.8 illustrates why we invert one of the functions and slide it left to right to do graphical convolution.

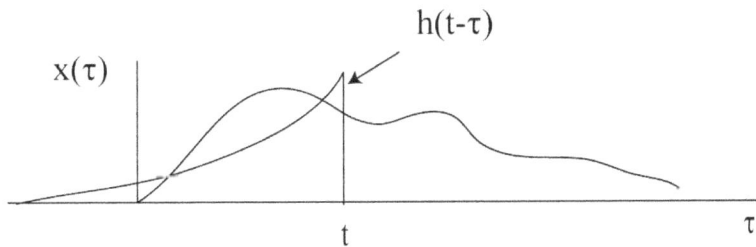

Figure 2.8. The graph of $x(\tau) h(t - \tau)$.

Figure 2.9 illustrates the convolution of $h(t) = e^{-0.2t}u(t)$ with $x(t) = u(t) - u(t-5)$.

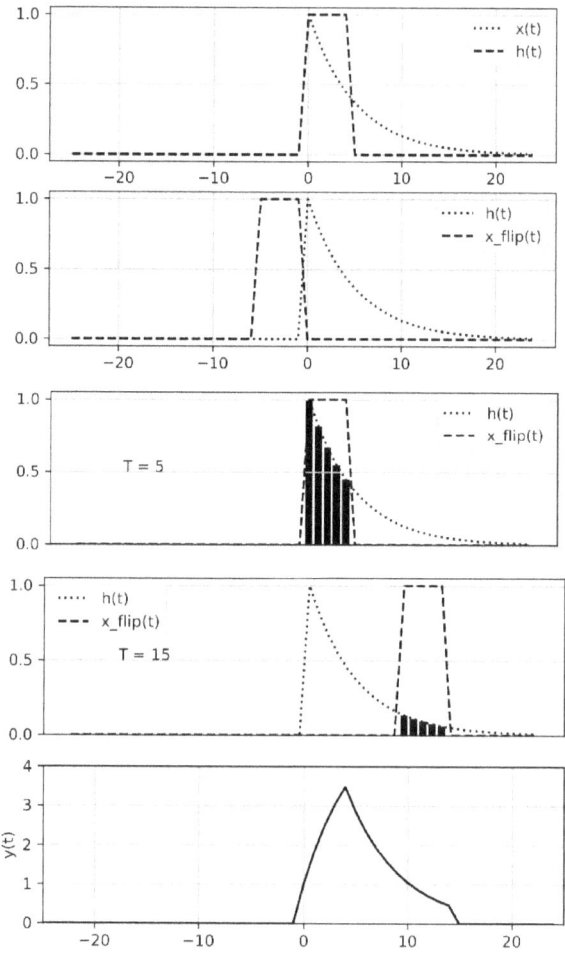

Figure 2.9. Convolution of $h(t) = e^{-0.2t}u(t)$ with $x(t) = u(t) - u(t-5)$.

Example 2.1. Solve the following differential equation for v_c:

$$\frac{dv_c(t)}{dt} + \frac{1}{RC}v_{out}(t) = u(t).$$

Solution

Since we already have the impulse response, the solution for the forcing function

$$x_{in}(t) = u(t)$$

is

$$v_{out}(t) = \int_0^t h(\tau) x_{in}(t-\tau)\, d\tau$$

$$= \int_0^t e^{-\tau/RC}\, d\tau.$$

Notice that it has been reduced to a simple integration when the input is a step function.

$$v_{out}(t) = \int_0^t e^{-\tau/RC}\, d\tau = (-RC) e^{-\tau/RC} \Big|_0^t = RC\left(1 - e^{-t/RC}\right) u(t).$$

The $u(t)$ is added on the end because clearly $v_{output}(t)$ has no value before $t = 0$.

Check

$$\frac{d}{dt} v_{out}(t) = RC\left(\frac{-1}{RC}\right)\left(-e^{-t/RC}\right) u(t) = e^{-t/RC} u(t)$$

So the original equation is

$$e^{-t/RC} u(t) + \frac{1}{RC} RC\left(1 - e^{-t/RC}\right) u(t) = u(t).$$

2.2.3 System Analysis Using the Impulse Response

Here, then, is a summary of the approach we have developed to analyze a system.

1. Use physical properties to develop a differential equation that relates output to input.
2. In the differential equation, use $\delta(t)$ as the forcing function, to get a solution called the impulse response, $h(t)$.
3. Now the response to any input $x(t)$ can be found by convolving $h(t)$ and $x(t)$.

Example 2.2. Suppose we want to mathematically describe the following situation. There is a vat whose total water content is described by v(t). The water going into the vat is described by f(t). However, there is a loss that is proportional to the volume already in the vat, given by 0.1v(t). Determine the volume of water as a function of time for the following cases:

(a) $f(t) = 10^2 u(t)$
(b) $f(t) = 10^2 e^{-0.5t} u(t)$

Solution

The rate of change in volume is
$$\frac{dv(t)}{dt} = f(t) - 0.1v(t),$$
from which we get the differential equation
$$\frac{dv(t)}{dt} + 0.1v(t) = f(t).$$
We want the impulse response, so we solve
$$\frac{dv(t)}{dt} + 0.1v(t) = \delta(t)$$
This is the same form as Eq. (2.4) above, so the impulse response must be
$$h(t) = e^{-0.1t} u(t).$$
The response to input (a) is
$$v(t) = \int_{-\infty}^{\infty} h(\tau) f(t-\tau)\, d\tau$$
$$= \int_{-\infty}^{\infty} 10^2 e^{-0.1\tau} u(\tau)\, u(t-\tau)\, d\tau.$$

The $u(\tau)$ means that we can start the integration at zero instead of infinity, and $u(t-\tau)$ means we need only take the integration up to t

instead of infinity, so it simplifies to

$$v(t) = \int_0^t 10^2 e^{-0.1\tau} d\tau = 10^2 \left(\frac{1}{-.1}\right)\left[e^{-0.1\tau}\right]_0^t$$

$$= 10^2 \left(\frac{1}{-.1}\right)\left[e^{-0.1t} - 1\right]u(t) = 10^3 \left(1 - e^{-0.1t}\right)u(t)$$

The $u(t)$ is added on the end because $v(t)$ is only defined starting at $t = 0$.
It does not matter which function we flip. We can also write

$$v(t) = \int_{-\infty}^{\infty} 10^2 e^{-0.1(t-\tau)} u(t-\tau) u(\tau) \, d\tau$$

$$= \int_0^t 10^2 e^{-0.1(t-\tau)} \, d\tau = 10^2 e^{-0.1t} \int_0^t e^{0.1\tau} \, d\tau$$

$$= 10^2 e^{-0.1t} \left(\frac{1}{0.1}\right)\left[e^{0.1\tau}\right]_0^t$$

$$= 10^3 e^{-0.1t} \left[e^{0.1t} - 1\right] = 10^3 \left(1 - e^{-0.1t}\right) u(t).$$

The response to input (b) is

$$v(t) = \int_0^t 10^2 e^{-0.1(t-\tau)} e^{-0.5\tau} \, d\tau$$

$$= 10^2 e^{-0.1t} \int_0^t e^{-0.4\tau} \, d\tau$$

$$= 10^2 e^{-0.1t} \left(\frac{1}{-0.4}\right)\left[e^{-0.4\tau}\right]_0^t$$

$$= -250 e^{-0.1t} \left[e^{-0.4t} - 1\right] = 250 \left(e^{-0.1t} - e^{-0.5t}\right) u(t).$$

We could have flipped this one around to get

$$v(t) = \int_0^t 10^2 e^{-0.1\tau} e^{-0.5(t-\tau)} \, d\tau = 10^2 e^{-0.5t} \int_0^t e^{0.4\tau} \, d\tau$$

$$= 10^2 e^{-0.5t} \left(\frac{1}{0.4}\right)\left[e^{0.4\tau}\right]_0^t$$

$$= 250 e^{-0.5t} \left[e^{0.4t} - 1\right] u(t) = 250 \left(e^{-0.1t} - e^{-0.5t}\right) u(t).$$

Note: Any physical system with a loss rate proportional to the total entity will invariably give an impulse response with an exponentially decaying term.

Example. 2.3. In a semiconductor, the number of excess electrons is given by $n(t)\ cm^{-3}$. Excess electrons are generated from a light source at a rate $G\ cm^{-3}s^{-1}$. However, if there are excess electrons, they will recombine with the excess holes that were also created at a rate given by

$$R = \frac{n(t)}{\tau_n}\ cm^{-3}s^{-1}$$

where $\tau_n = 10^{-6}s$ is the lifetime of the electron.
What is $n(t)$ for the following light sources?:

(a) $G(t) = 10^8 u(t)$
(b) $G(t) = 10^8 \sin(10^6 t)\quad -\infty < t < \infty$.

Solution

We start by finding a differential equation to describe this system, from which we can determine an impulse response. The rate of change of $n(t)$ is the rate at which electrons are generated minus the rate at which they recombine,

$$\frac{dn(t)}{dt} = G(t) - \frac{n(t)}{\tau_n},$$

or

$$\frac{dn(t)}{dt} + \frac{n(t)}{\tau_n} = G(t).$$

To find the impulse response, we solve for $n(t)$ using $\delta(t)$

$$\frac{dn(t)}{dt} + \frac{n(t)}{\tau_n} = \delta(t).$$

The mathematics to find the impulse response is now the same as above

so we can just write
$$h(t) = e^{-t/\tau_n} u(t),$$
and therefore, the solution for the input (a) will be
$$n(t) = e^{-t/10^{-6}} u(t) * 10^8 u(t).$$
Since both h(t) and G(t) are causal functions, we can write
$$n(t) = \int_0^t G(\tau) h(t-\tau) \, d\tau = \int_0^t 10^8 e^{-\left(\frac{t-\tau}{10^{-6}}\right)} d\tau$$
$$= 10^8 e^{-(10^6 t)} \int_0^t e^{10^6 \tau} \, d\tau = 10^8 e^{-10^6 t} \left(\frac{1}{10^6}\right) \left[e^{10^6 \tau}\right]_0^t$$
$$= 10^2 \left(1 - e^{-10^6 t}\right) u(t).$$

What if we switched G(t) and h(t) around?
$$n(t) = \int_0^t G(t-\tau) h(\tau) \, d\tau = \int_0^t 10^8 e^{-\left(\frac{\tau}{10^{-6}}\right)} d\tau$$
$$= 10^8 \left(\frac{1}{10^6}\right) \left[e^{10^6 \tau}\right]_0^t = 10^2 \left(1 - e^{-10^6 t}\right) u(t).$$

The response to (b) will be a different story because the input is not causal
$$n(t) = \int_{-\infty}^{\infty} h(t-\tau) G(\tau) \, d\tau$$
$$= \int_{-\infty}^{\infty} e^{-10^6 (t-\tau)} u(t-\tau) 10^8 \sin(10^6 \tau) d\tau$$
$$= 10^8 e^{-10^6 t} \int_{-\infty}^{t} e^{10^6 \tau} \sin(10^6 \tau) d\tau$$
$$= 10^8 e^{-10^6 t} \frac{e^{10^6 t} \left(10^{-6} \sin(10^6 t) - 10^6 \cos(10^6 t)\right)}{10^{12} + 10^{12}}$$
$$= \frac{10^2}{2} \left(\sin(10^6 t) - \cos(10^6 t)\right)$$

Note that this is not causal. Of course, it is much easier to do it using

45

phasors. The phasor equation is
$$j\omega N + 10^6 N = 1$$
$$N = \frac{1}{j10^6 + 10^6} = \frac{10^{-6}}{1+j} = \frac{10^{-6}}{\sqrt{2}\angle 45^\circ}$$
From which we get the answer
$$n(t) = \frac{10^8 10^{-6}}{\sqrt{2}} \sin\left(10^6 t - 45^\circ\right).$$

We will see that the response to (a) can be more easily calculated with Laplace transforms.

Example 2.4. Find the impulse response of the following second order system:
$$\frac{d^2 y(t)}{dt^2} + 4\frac{dy(t)}{dt} + 3y(t) = \delta(t). \quad (2.17)$$

Solution

The characteristic equation is
$$s^2 + 4s + 3 = (s+3)(s+1) = 0$$
so the homogenous solution will be of the form
$$y(t) = \left(Ae^{-3t} + Be^{-t}\right)u(t). \quad (2.18\ \text{a})$$
The first derivative is
$$\frac{y(t)}{dt} = \left(-3Ae^{-3t} - Be^{-t}\right)u(t) + (A+B)\delta(t) \quad (2.18\ \text{b})$$
and the second derivative is
$$\frac{y^2(t)}{dt^2} = \left(9Ae^{-3t} + Be^{-t}\right)u(t)$$
$$+ (-3A - B)\delta(t) + (A+B)\delta^{(1)}(t) \quad (2.18\ \text{c})$$
Putting these back into Eq. (2.1.13) gives

$$(9Ae^{-3t} + Be^{-t})u(t) + (-3A - B)\delta(t) + (A + B)\delta^{(1)}(t)$$
$$+ 4\left[(-3Ae^{-3t} - Be^{-t})u(t) + (A + B)\delta(t)\right]$$
$$+ 3\left[(Ae^{-3t} + Be^{-t})u(t)\right] = \delta(t).$$

If Eq. (2.17) is to hold true, then the coefficients for the different types of functions must be satisfied, so we get three equations:

The $\delta^{(1)}(t)$ terms give $\quad (A + B)\delta^{(1)}(t) = 0$
or $\quad A = -B$.
The $\delta(t)$ terms give $(-3A - B)\delta(t) + 4(A + B)\delta(t) = \delta(t)$
or $\quad (-3A + A) = 1 \Rightarrow A = -\dfrac{1}{2}, \quad B = \dfrac{1}{2}.$

The $u(t)$ terms give
$$(9Ae^{-3t} + Be^{-t}) + 4(-3Ae^{-3t} - Be^{-t}) + 3(Ae^{-3t} + Be^{-t}) = 0,$$
But this is redundant, because our choice of the homogeneous equation insured it. Therefore, we can conclude,
$$h(t) = \frac{1}{2}(e^{-t} - e^{-3t})u(t).$$

What would be the response for the input $x(t) = u(t)$?
$$y(t) = h(t) * u(t) = \int_0^t h(\tau)\, d\tau$$
$$= \int_0^t \frac{1}{2}(e^{-\tau} - e^{-3\tau})\, d\tau = \frac{1}{2}\left\{(1 - e^{-t}) - \frac{1}{3}(1 - e^{-3t})\right\}u(t)$$
$$= \frac{1}{2}(e^{-3t} - e^{-t})u(t).$$

This problem was not difficult because the characteristic equation separated into two simple real roots. In general, it will be much easier to solve differential equations like this by using Laplace transforms.

2.3 Convolution Properties

There are several important properties associated with convolution, many of which are obvious by the graphical methods. They can also be proven just from the mathematical definition. The formal definition of convolution is

$$y(t) = h(t) * x(t) = \int_{-\infty}^{\infty} h(t-\tau)x(\tau)\, d\tau, \qquad (2.19)$$

but we have already said that if both $h(t)$ and $x(t)$ are causal, Eq. (2.19) can be written,

$$h(t) * x(t) = \int_{0}^{t} h(\tau)x(t-\tau)\, d\tau. \qquad (2.20)$$

The following are properties related to convolution:

1. <u>Commutative Property</u>

$$\begin{aligned} h(t) * x(t) &= x(t) * h(t) \\ &= \int_{0}^{t} h(t-\tau)x(\tau)\, d\tau = \int_{0}^{t} x(t-\tau)h(\tau)\, d\tau \end{aligned} \qquad (2.21)$$

This one is easy to prove mathematically. Start with a change of variables

$$\sigma = t - \tau. \qquad (2.22)$$

This implies,

$$\tau = t - \sigma,$$
$$d\tau = -d\sigma.$$

Notice also how the limits of integration change when using Eq. (2.22),

$$\tau = 0 \quad \rightarrow \quad \sigma = t,$$
$$\tau = t \quad \rightarrow \quad \sigma = 0,$$

so the first integral in Eq. (2.21) becomes,

$$\int_{0}^{t} h(t-\tau)x(\tau)\, d\tau = \int_{t}^{0} h(\sigma)x(t-\sigma)\, -d\sigma$$
$$= \int_{0}^{t} h(\sigma)x(t-\sigma)\, d\sigma,$$

which is the same as the second integral in Eq. (2.21), except the dummy variable is σ instead of τ. This is easy to see graphically. Suppose we have two functions in Fig. 2.21. If we graphically do the convolution indicated by the first integral of Eq. (2.21), we invert the *h(t)* function around the origin and slide it to the right to find the overlap of the two areas as a function of time. If we use the second integral of Eq. (2.21), we first invert the *x(t)* function and move it. Obviously, both procedures result in the same graph.

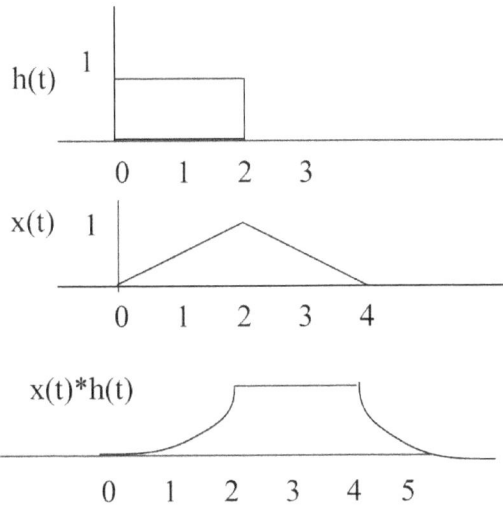

Figure 2.10. Graphical convolution.

2. <u>Distributive Property</u>

This property says that for three functions

$$f_1(t) * \{f_2(t) + f_3(t)\} = f_1(t) * f_2(t) + f_1(t) * f_3(t) \qquad (2.23)$$

The mathematical proof is trivial because this is just a statement of linearity. However, this property is very useful. Look at the convolution in Fig. 2.11. By far, the easiest way to do this convolution is to break *h(t)* into two separate functions, use convolution with these separate pieces, and then add them together at the end.

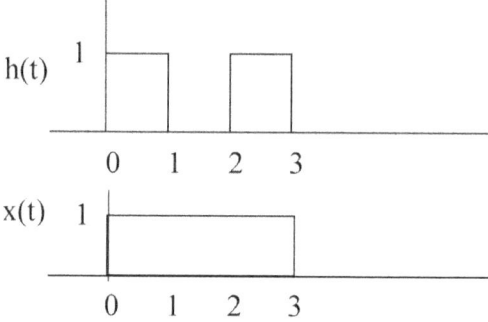

(a) The two functions to be convolved

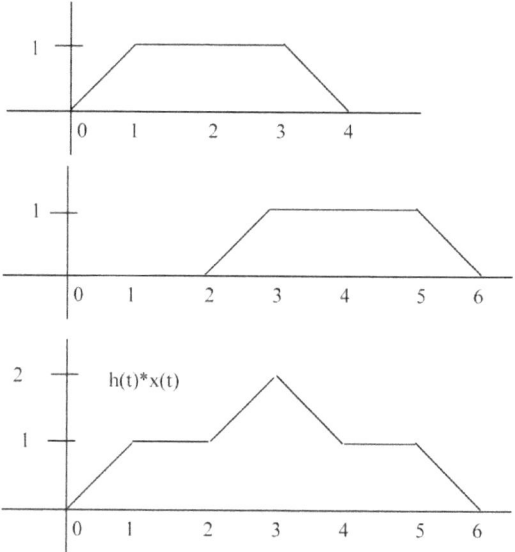

(b) The two parts of *h(t)* are convolved separately with x(t) and then added together.

Figure 2.11. Graphical convolution using distribution.

In Fig. 2.11a, the top graph is the convolution of *h(t)* with the first part of *x(t)* and the second graph is the convolution of *h(t)* with the second part of *x(t)*. The bottom is the sum of the two.

3 <u>Associative Property</u>

This simply says that when convolving three or more functions, the order does not matter.

$$f_1(t) * \{f_2(t) * f_3(t)\} = \{f_1(t) * f_2(t)\} * f_3(t) \quad (2.24)$$

This involves convolution with double integrals, which is outside the scope of this book. This is what it looks like:

$$f_2(t) * f_3(t) = \int_0^t f_2(t-\tau) f_3(\tau) \, d\tau = F(t).$$

So now Eq. (2.24) becomes,

$$f_1(t) * F(t) = \int_0^t f_1(t-\tau') F(\tau') \, d\tau', \quad (2.25)$$

where τ' is just a different dummy variable. Now Eq. (2.25) becomes

$$f_1(t) * \{f_2(t) * f_3(t)\} = \int_0^t f_1(t-\tau') \int_0^{\tau'} f_2(\tau'-\tau) f_3(\tau) \, d\tau \, d\tau'$$

$$= \int_0^t f_1(t-\tau') \int_0^{\tau'} f_2(\tau) f_3(\tau'-\tau) \, d\tau \, d\tau'.$$

4. Duration Property

If $f_1(t)$ only exists on the interval $[t_1, T_1]$, and $f_2(t)$ only exists on the interval $[t_2, T_2]$, then the convolution integral is only defined for the interval $[t_1 + t_2, T_1 + T_2]$, i.e.,

$$f_1(t) * f_2(t) = \int_{t_1+t_2}^{T_1+T_2} f_1(t-\tau) f_2(\tau) \, d\tau.$$

This is particularly useful in doing graphical convolution.

5. Time-Shifting Property

If $y(t) = h(t) * x(t)$,

then $y(t-\sigma) = h(t-\sigma) * x(t)$

and $y(t-\sigma_1-\sigma_2) = h(t-\sigma) * x(t-\sigma_2)$

2.3.1 Special Convolutions

Two functions whose convolutions yield special results are worth noting.

Delta function $\delta(t)$:

$$f(t) * \delta(t) = \int_{-\infty}^{\infty} f(\tau)\delta(t-\tau)\,d\tau = f(t) \qquad (2.26)$$

The integral in Eq. (2.26) only has a value when $t-\tau$, so in effect, the δ just traces out the function.

Unit step function $u(t)$:

$$f(t) * u(t) = \int_{-\infty}^{\infty} f(\tau)u(t-\tau)\,d\tau$$

$$= \int_{-\infty}^{t} f(\tau)u(t-\tau)\,d\tau \qquad (2.27)$$

$$= \int_{-\infty}^{t} f(\tau)\,d\tau$$

The upper limit is changed to t because the "flipped" function $u(t)$ only has a value up to t. In the second integral, $u(t-\tau)$ can just be replaced with one.

2.3.2 Graphical Convolution

Sometimes, convolution can most easily be performed graphically. The following steps are a guide to graphical convolution.

Step 1 Use the duration property, #4 above, to find the interval on which the integral is non-zero.

Step 2 Flip one signal about the vertical axis (usually the simpler one).

Step 3 Vary the parameter τ looking for the overlap.

Exercises

2.1 A system is described by the following equation:

$$\frac{dy(t)}{dt} + 20y(t) = f(t),$$

where $y(t)$ is the output and $f(t)$ is the input.

a. What is the impulse response of this system?

b. What is the response to $f(t) = u(t) - u(t-0.1)$?

c. What is the response to $f(t) = 3\cos(10t - 30°)$?

2.2. In the following problem, use the methods from this chapter. Do not use Laplace transforms. For the following RL circuit $R = 100\ k\Omega$ and $L = 1\ \mu H$. At time time $t = 0$ $i(0^-) = 0\ A$.

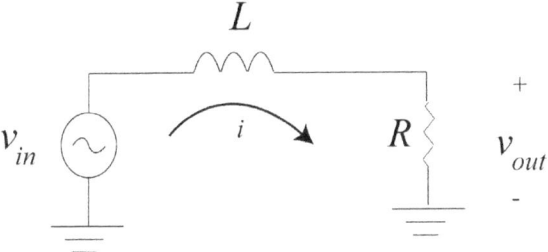

a. Write a differential equation describing this circuit with v_{in} as the input and v_{out}

b. What is the impulse response?

c. What is the response to the input $v_{in} = u(t)$?

d. What is the response if $v_{in} = e^{-5t}u(t)$?

2.3. A system is described by the following differential equation:
$$\frac{dy(t)}{dt} + 0.3y(t) = f(t).$$
(a) What is the impulse response to this system?

(b) If $f(t) = \delta(t) + \delta(t-1) + \delta(t-2)$, what is $y(t)$?

(c) If $f(t) = u(t)$ what is $y(t)$?

(d) Find $y(t)$ if $f(t) = 4\cos(3t)$ $-\infty < t < \infty$

2.4. The impulse response of a system is given by
$$h(t) = (e^{-t} - e^{-2t})u(t).$$
Determine the response to the following input:
$$x(t) = u(t) - u(t-2).$$

2.5. It is found that the total number of bacteria in a dish, $n(t)$ grow in proportion to a light source shining on the dish. This rate is $I = 10^6$ min^{-1}. However, at any given time, five percent of the total bacteria die per minute.

a. Write a differential equation describing the total number of bacteria if the light source is turned on at time t=0. (At $t = 0$, assume $n = 0$.)

b. If we were to model this as a system with light being the input and the bacteria population being the output, what would be the impulse response?

2.6. Graphically convolve the following two functions.

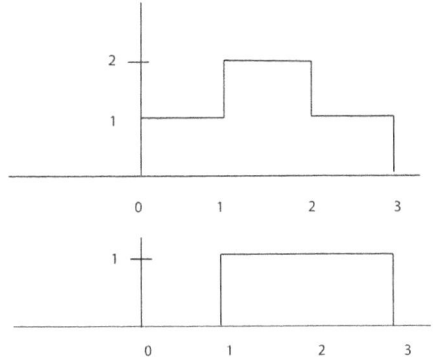

2.7. Graphically convolve the following two functions.

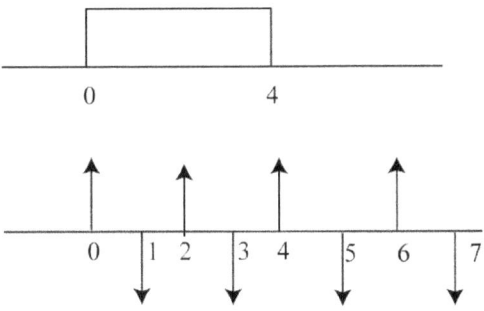

Chapter 3. Laplace Transforms

In the two previous Chapters we talked about the solutions to differential equations with causal forcing functions, i.e., functions that begin at time zero. In this Chapter we introduce the Laplace transform, which is the best approach for solving this type of equation.

3.1. Introduction to the Laplace Transform

The Laplace transform of a time-domain signal $f(t)$ is given by

$$F(s) = \mathcal{L}\{f(t)\} = \int_{0^-}^{\infty} f(t) e^{-st} dt . \quad (3.1)$$

Notice that the lower limit of the integral is 0^-, which means, just before zero. This is done to indicate that the integral includes zero. As an example, the Laplace transform of an exponential function is

$$\begin{aligned} F(s) &= \mathcal{L}\{e^{-\alpha t} u(t)\} \\ &= \int_{0^-}^{\infty} e^{-\alpha t} u(t) e^{-st} dt = \int_{0^-}^{\infty} e^{-(s+\alpha)t} dt \\ &= \frac{1}{-(s+\alpha)} e^{-(s+\alpha)t} \Big|_{0^-}^{\infty} = \frac{0-1}{-(s+\alpha)} \\ &= \frac{1}{(s+\alpha)} . \end{aligned} \quad (3.2)$$

Notice that

$$e^{-\alpha t} u(t) = u(t) \quad \text{if } \alpha = 0 ,$$

so we can conclude

$$\mathcal{L}\{u(t)\} = \frac{1}{s} . \quad (3.3)$$

Here is an interesting transform:

$$\mathcal{L}\{\delta(t)\} = \int_{0^-}^{\infty} \delta(t) e^{-st} dt = 1 . \quad (3.4)$$

The main reason for using 0^- as the lower limit is that the delta function is completely included in the integration. Throughout the rest of the chapter, it should be understood that the lower limit is 0^-, even if it is not stated explicitly.

We know the Laplace transform of f(t), but since we want to use it to solve differential equations, we will also want to know the Laplace transform of the derivative.

Property 1. Derivative

If $F(s)$ is the Laplace transform of f(t), then

$$\mathcal{L}\left\{\frac{df(t)}{dt}\right\} = sF(s) - f(0^-), \qquad (3.5)$$

Where $f(0^-)$ is the value of f(t) at time $t = 0^-$. Before we prove this equation, recall the method of *integration by parts:*

$$\int_a^b \frac{df(t)}{dt} g(t)\,dt = f(t)g(t)\Big|_a^b - \int_a^b f(t)\frac{dg(t)}{dt}\,dt.$$

Proof

$$\mathcal{L}\left\{\frac{df(t)}{dt}\right\} = \int_{0^-}^{\infty} \frac{df(t)}{dt} e^{-st}\,dt$$

$$= f(t)e^{-st}\Big|_{0^-}^{\infty} - \int_{0^-}^{\infty} f(t)\cdot-se^{-st}\,dt$$

$$= -f(0^-) + s\int_{0^-}^{\infty} f(t)e^{-st}\,dt$$

$$= sF(s) - f(0).$$

This can be generalized to higher orders, e.g.,

$$\mathcal{L}\left\{\frac{d^2 f(t)}{dt^2}\right\} = s^2 F(s) - sf(0) - f'(0)$$

The term $f(0^-)$ is an *initial condition*. It is the value of f(t) at time

$t = 0^-$, i.e., the time just before zero. We will continue to indicate all initial conditions as $f(0)$, but it should be understood to mean $f(0^-)$.

With just these few transforms and the derivative property, let us see what we can do. Suppose we want the impulse response of the first order differential equation given by

$$\frac{dy(t)}{dt} + 0.3y(t) = \delta(t). \tag{3.6}$$

The Laplace transform of this equation is just the transform of the individual components, so

$$sY(s) - y(0) + 0.3Y(s) = 1.$$

In determining the impulse response, initial conditions play no role, so we may assume $y(0) = 0$. This leaves

$$Y(s) = \frac{1}{s + 0.3}.$$

We do not know how to take inverse Laplace transforms yet, but we know from Eq. (3.2) that the time domain version of this function, which is our impulse response, is

$$h(t) = e^{-0.3t}u(t). \tag{3.7}$$

Now suppose we take our original equation and say that the input is the step function.

$$\frac{dy(t)}{dt} + 0.3y(t) = u(t). \tag{3.8}$$

Since we have the impulse response, we can convolve $u(t)$ with $h(t)$. However, let us just take the Laplace transform of the original equation with the new forcing function and see what happens:

$$sY(s) + 0.3Y(s) = \frac{1}{s}, \tag{3.9}$$

where we have assumed the system was at rest. Solving Eq. (3.9) for $Y(s)$,

$$Y(s) = \frac{1}{s + 0.3} \cdot \frac{1}{s}. \tag{3.10}$$

We do not specifically know the inverse of this function, but we do know the inverses of the two separate functions on the right side of the equation, so we separate them by *partial fraction expansion*:

$$Y(s) = \frac{1}{s+0.3} \cdot \frac{1}{s} = \frac{A}{s+0.3} + \frac{B}{s} \qquad (3.11)$$

There are several ways to solve this for A and B. The most straightforward is to simply multiply the two terms to get the original

$$Y(s) = \frac{1}{s+0.3} \cdot \frac{1}{s} = \frac{A}{s+0.3} + \frac{B}{s} = \frac{As + B(s+0.3)}{(s+0.3)s}.$$

For this to be true, all the terms in the numerator must add up to one, giving the two equations

$$s \text{ terms}: \ 0 = A + B,$$
$$s^0 \text{ terms}: 1 = 0.3B,$$

from which we conclude

$$B = \frac{1}{.3} = 3.33, \quad A = -B = -3.33,$$

$$Y(s) = 3.333\left(\frac{1}{s} - \frac{1}{s+0.3}\right).$$

Taking the inverse of the two s-domain functions gives the following time-domain function

$$y(t) = 3.333\left(1 - e^{-0.3t}\right)u(t). \qquad (3.12)$$

This is pretty interesting; we solved this without doing a convolution.

At this point, we are going to stop and look at one of the most important properties of Laplace transforms.

Property 2. Convolution

If $f(t)$ and $h(t)$ are both causal functions, then the Laplace Transform of the convolution of $f(t)$ and $h(t)$ is simply the product of the Laplace transforms of the individual functions:

$$\mathcal{L}\{f(t)*h(t)\} = F(s)H(s). \qquad (3.13)$$

Proof

$$\mathcal{L}\{f(t)*h(t)\} = \int_0^\infty \left[\int_0^t f(t-\tau)h(\tau)d\tau\right] e^{-st} dt.$$

(Note: the function in the brackets is a convolution. In doing the convolution of this function, we have to evaluate it at infinity, so we will change the upper limit of the integration to infinity.) We start by changing the order of integration,

$$\mathcal{L}\{f(t)*h(t)\} = \int_0^\infty h(\tau)\left[\int_0^\infty f(t-\tau)e^{-st} dt\right] d\tau.$$

Now we change variables, using

$$\sigma = t - \tau, \quad dt = d\sigma.$$

$$\int_0^\infty h(\tau)\left[\int_0^\infty f(\sigma)e^{-s(\sigma+\tau)} d\sigma\right] d\tau = \int_0^\infty h(\tau)e^{-s\tau} d\tau \left[\int_0^\infty f(\sigma)e^{-s\sigma} d\sigma\right]$$

$$= H(s)F(s).$$

Can this property be true? Instead of doing the convolution, can we take the Laplace transforms of the two functions and just multiply them? Yes, it is true. With these simple examples using just a few functions, we have demonstrated the two main reasons for using Laplace transforms:

1. Differential equations become algebraic equations.
2. Convolution becomes multiplication.

Example 3.1. An RC circuit has an impulse response given by
$$h(t) = e^{-5t}u(t).$$
What is the time-domain response to a unit step function, *f(t) = u(t)*?

Solution

The time domain response is the convolution of the impulse response and the input

$$y(t) = \int_0^t h(t-\tau)f(\tau)d\tau.$$

However, we will use the convolution property, Eq. (3.13). The Laplace transform of the impulse response is

$$H(s) = \mathcal{L}\left\{e^{-5t}u(t)\right\} = \frac{1}{s+5}.$$

The Laplace transform of the input is

$$F(s) = \mathcal{L}\left\{u(t)\right\} = \frac{1}{s}.$$

The output in the Laplace domain is

$$Y(s) = H(s)F(s) = \frac{1}{s+5}\frac{1}{s} = \frac{1/5}{s} - \frac{1/5}{s+5}.$$

Therefore, the time domain response is

$$y(t) = \frac{1}{5}\left(1-e^{-5t}\right)u(t).$$

Example 3.2. For the same impulse response in Example 3.1.1, find the response to the following input:

$$f(t) = e^{-2t}u(t).$$

Solution

$$F(s) = \mathcal{L}\left\{e^{-2t}u(t)\right\} = \frac{1}{s+2},$$

$$Y(s) = \frac{1}{s+5}\frac{1}{s+2} = \frac{-1/3}{s+5} + \frac{1/3}{s+2}.$$

The inverse Laplace of $Y(s)$ is

$$y(t) = \frac{1}{3}\left(e^{-2t} - 3^{-5t}\right)u(t).$$

3.1.1. Partial Fraction Expansion

Look the following differential equation,

$$\frac{dy(t)}{dt} + 0.3y(t) = e^{-0.1t}u(t) \quad y(0) = 0. \tag{3.14}$$

First take the Laplace transform of both sides

$$sY(s) + 0.3Y(s) = \frac{1}{s+0.1},$$

and then solve for $Y(s)$

$$Y(s) = \frac{1}{s+0.3}\frac{1}{s+0.1}.$$

Ultimately, we always want to go back to the time domain. There is an inverse Laplace transform given by

$$f(t) = \frac{1}{2\pi j}\int_{-j\infty}^{j\infty} F(s)e^{st}ds.$$

These are solved by techniques referred to as *contour integration*. As you have probably noticed, our approach is simply to get the Laplace-domain function in a form we recognize, and take it back to the time domain, term by term. Once again we write

$$Y(s) = \frac{1}{s+0.3}\frac{1}{s+0.1} = \frac{A}{s+0.3} + \frac{B}{s+0.1}. \tag{3.15}$$

This time, we will be a little more systematic. For A and B to be the correct values, they have to be the correct values *everywhere*. We will exploit this. To look for A, multiply through by $s+0.3$

$$Y(s)(s+0.3) = \frac{1}{s+0.1} = A + \frac{B}{s+0.1}(s+0.3).$$

In order to isolate A, we evaluate the entire equation at $s = -0.3$:

$$\left.\frac{1}{s+0.1}\right|_{s=-0.3} = A + \left.\frac{B}{s+0.1}(s+0.3)\right|_{s=-0.3}.$$

The term containing B falls out, and we are left with,

$$A = \left.\frac{1}{s+0.1}\right|_{s=-0.3} = \frac{1}{-.3+0.1} = \frac{1}{-.2} = -5.$$

A similar procedure gives us B:

$$B = \frac{1}{s+0.3}\bigg|_{s=-0.1} = \frac{1}{-.1+0.3} = \frac{1}{.2} = 5.$$

Equation (3.15) then becomes

$$Y(s) = \frac{-5}{s+0.3} + \frac{5}{s+0.1}. \tag{3.16}$$

The corresponding time domain function is,

$$y(t) = 5\left(e^{-0.1t} - e^{-0.3t}\right)u(t). \tag{3.17}$$

Now go back to the original problem, but this time, assume we have an initial condition of $y(0) = 0.2$, i.e,

$$\frac{dy(t)}{dt} + 0.3y(t) = e^{-0.1t}u(t), \quad y(0^-) = 0.2. \tag{3.18}$$

Now when we take the Laplace transform of Eq. (3.18) we use Eq. (3.5) to get,

$$sY(s) - y(0) + 0.3Y(s) = \frac{1}{s+0.1}.$$

Solving for Y(s) gives

$$Y(s) = \frac{0.2}{s+0.3} + \frac{1}{s+0.3}\frac{1}{s+0.1},$$

We already know the inverse of the second term on the right. The inverse of the term containing the initial condition just adds an exponential term:

$$y(t) = 0.2e^{-0.3t}u(t) + 5\left(e^{-0.1t} - e^{-0.3t}\right)u(t)$$
$$= \left(5e^{-0.1t} - 4.8e^{-0.3t}\right)u(t). \tag{3.19}$$

This demonstrates the full power of the Laplace transform method for solving differential equations. The initial condition just looks like another input.

Example 3.3 Find the inverse Laplace transform of

$$F(s) = \frac{s}{s+3}. \tag{3.20}$$

Solution

This looks easy enough. However, when we look at a table of Laplace transforms (such as Table 3.1), we do not find this transform in the table. This is because the partial fraction expansion cannot be used directly on an equation like Eq. (3.2). The order of the numerator must be one less than the denominator. So instead we just divide

$$F(s) = \frac{s}{s+3} = 1 - \frac{3}{s+3}.$$

The inverse Laplace of this is

$$f(t) = \delta(t) - 3e^{-3t}u(t).$$

What if we looked at it this way:

$$F(s) = s\left[\frac{1}{s+3}\right] \qquad (3.21)$$

It almost looks like we could use the derivative rule,

$$\mathcal{L}\left[\frac{d}{dt}f(t)\right] = sF(s) - f(0^-).$$

What is $f(0^-)$? You might be tempted to say

$$f(0) = e^{-3t}u(t)\Big|_{t=0} = 1.$$

But be careful, $f(0^-)$ means the value of $f(t)$ right before time $t=0$. Therefore,

$$f(0^-) = e^{-3t}u(t)\Big|_{t=0^-} = 0.$$

This will be true for any causal function. So if $f(t)$ is a causal function and its Laplace transform is $F(s)$,

$$\mathcal{L}\left[\frac{d}{dt}f(t)\right] = sF(s).$$

So we could say we have a theorem that remains true when $f(0^-) = 0$:

$$\mathcal{L}^{-1}\left[sF(s)\right] = \frac{d}{dt}f(t).$$

We can use the derivative rule for this problem:

$$\mathcal{L}^{-1}\left[\frac{s}{s+3}\right] = \mathcal{L}^{-1}\left[s\left(\frac{1}{s+3}\right)\right]$$

$$= \frac{d}{dt}\left(e^{-3t}u(t)\right) = -3e^{-3t}u(t) + \delta(t).$$

Before going on, we will find the Laplace transform of a new function, $f(t) = te^{-\alpha t}u(t)$. To do so, we just use the definition,

$$\mathcal{L}\{te^{-\alpha t}u(t)\}dt = \int_0^\infty te^{-\alpha t}e^{-st}dt.$$

This is not an integral that most of us remember, but it is readily available in most basic tables of integrals. A table of indefinite integrals informs us that

$$\int xe^{ax}dx = \frac{e^{ax}}{a}\left(x - \frac{1}{a}\right).$$

Therefore, the above integral becomes

$$\int_0^\infty te^{-\alpha t}e^{-st}dt = \int_0^\infty te^{-(s+\alpha)t}dt = \left.\frac{e^{-(s+\alpha)t}}{-(s+\alpha)}\left(t - \frac{1}{-(s+\alpha)}\right)\right|_{t=0}^{t=\infty} \quad (3.22)$$

$$= 0 - \left(\frac{1}{-(s+\alpha)}\right)\left(0 - \frac{1}{-(s+\alpha)}\right) = \frac{1}{(s+\alpha)^2}.$$

Notice that e^{-st} tends to make the limit at $t = \infty$ converge to zero. This is one of the strengths of the Laplace transform and one of the key reasons it was developed.

More generally,

$$\mathcal{L}\{t^n e^{-\alpha t}u(t)\} = \frac{n!}{(s+\alpha)^{n+1}}. \quad (3.23)$$

Example 3.4 Find

$$\mathcal{L}^{-1}\left[\frac{s}{(s+3)^2}\right].$$

Solution

$$\mathcal{L}^{-1}\left[s\left(\frac{1}{(s+3)^2}\right)\right] = \frac{d}{dt}\left[te^{-3t}u(t)\right]$$
$$= e^{-3t}u(t) - 3te^{-3t}u(t) + te^{-3t}\delta(t)$$
$$= e^{-3t}u(t) - 3te^{-3t}u(t)$$

We can check our answer by taking the Laplace transforms of the individual time-domain terms:

$$\mathcal{L}\left\{e^{-3t}u(t) - 3te^{-3t}u(t)\right\} = \frac{1}{s+3} - 3\frac{1}{(s+3)^2}$$
$$= \frac{s+3-3}{(s+3)^2} = \frac{s}{(s+3)^2}.$$

3.2 Theorems and Properties of the Laplace Transform

We have already discovered some of the characteristics of the Laplace transform. Here we learn some more. The proofs stem directly from the definition of the Laplace transform.

Property 3. Linearity

The Laplace transform of the linear sum of two functions is the linear sum of their Laplace transforms:

$$\mathcal{L}\left\{\alpha f_1(t) + \beta f_2(t)\right\} = \alpha F_1(s) + \beta F_2(s).$$

Proof

The proof comes from the definition:

$$\mathcal{L}\{\alpha f_1(t)+\beta f_2(t)\}=\int_0^\infty [\alpha f_1(t)+\beta f_2(t)]e^{-st}dt$$
$$=\int_0^\infty \alpha f_1(t)e^{-st}dt+\int_0^\infty \beta f_2(t)e^{-st}dt \quad (3.24)$$
$$=\alpha F_1(s)+\beta F_2(s).$$

Property 4. Integration

$$\mathcal{L}\left\{\int_0^t f(\tau)d\tau\right\}=\frac{1}{s}F(s). \quad (3.25)$$

Proof

The easiest way to prove this property is to remember that the integration of a causal function may be thought of as convolution with the unit step function

$$\mathcal{L}\left\{\int_0^t f(\tau)d\tau\right\}=f(t)*u(t).$$

Now use the convolution theorem, Eq. (3.1.13):

$$\mathcal{L}\{f(t)*u(t)\}=\frac{1}{s}F(s). \quad (3.26)$$

Property 5. Time Scaling

There are times in engineering when it is desirable to use a different time scale on a function. The time scaling property can be useful in finding the Laplace transform of the scaled function.

$$\mathcal{L}\{f(at)\}=\frac{1}{a}F\left(\frac{s}{a}\right). \quad (3.27)$$

Proof

We start with Eq. (3.27) and change variables:

$$\mathcal{L}\{f(at)\} = \int_0^\infty f(at)e^{-st}\,dt \qquad \sigma = at$$
$$= \int_0^\infty f(\sigma)e^{-s(\sigma/a)}\frac{d\sigma}{a} = \frac{1}{a}\int_0^\infty f(\sigma)e^{-\sigma(s/a)}\,d\sigma$$
$$= \frac{1}{a}F\left(\frac{s}{a}\right).$$

The last step occurs because the integral looks like the Laplace transform, except s has been replaced by s/a. This is only valid for a greater than zero because Laplace is a causal transform.

Property 6. Time Shifting

$$\mathcal{L}\{f(t-t_0)u(t-t_0)\} = e^{-st_0}F(s) \qquad (3.28)$$

Proof

$$\mathcal{L}\{f(t-t_0)u(t-t_0)\} = \int_{t_0}^\infty f(t-t_0)e^{-st}\,dt.$$

Changing variables to $\tau = t - t_0$,

$$\mathcal{L}\{f(t-t_0)u(t-t_0)\} = \int_0^\infty f(\tau)e^{-s(\tau+t_0)}\,d\tau$$
$$= e^{-st_0}\int_0^\infty f(\tau)e^{-s\tau}\,d\tau = e^{-st_0}F(s).$$

Example 3.5 Solve the following:

$$\frac{dy(t)}{dt} + 0.2y(t) = e^{-0.5(t-1)}u(t-1) \quad y(0)=0.$$

Solution

Start by taking the Laplace transform of both sides.

$$sY(s) + 0.2Y(s) = \frac{e^{-s}}{s+0.5}.$$

Warning: The e^{-s} appears in the numerator because of the delay. However, do not try to incorporate that into the partial fraction expansion solution. Instead, set it aside and make the shift at the end. Using this procedure we solve

$$Y(s) = \left[\frac{1}{s+0.2}\frac{1}{s+0.5}\right]e^{-s} = \left[\frac{A}{s+0.2} + \frac{B}{s+0.5}\right]e^{-s}$$

$$= \left[\frac{3.33}{s+0.2} - \frac{3.33}{s+0.5}\right]e^{-s}.$$

Then when we go back to the time domain, we make the shift.

$$y(t) = 3.33\left(e^{-0.2(t-1)} - e^{-0.5(t-1)}\right)u(t-1).$$

Example 3.6 Find the Laplace transform of $f(t) = e^{-at}u(t-t_0)$.

Solution
Be careful. To use the time-shifting theorem the t terms all have to be in the same form.

$$f(t) = e^{-at_0}e^{-a(t-t_0)}u(t-t_0)$$

$$\mathcal{L}\{f(t)\} = e^{-at_0}\mathcal{L}\left\{e^{-a(t-t_0)}u(t-t_0)\right\}$$

$$= e^{-at_0}e^{-st_0}\frac{1}{s+a}.$$

Example 3.7 Find the Laplace transform of the following function:

$$f(t) = (t+1)e^{-t}u(t-1).$$

Solution

In order to use the time shifting theorem, every time term must be in the same form as the step function:

$$e^{-t} = e^{-1}e^{-(t-1)},$$

$$(t+1) = (t-1) + 2.$$

Now we have

$$f(t) = \left[(t-1) + 2\right]e^{-1}e^{-(t-1)}u(t-1)$$
$$= e^{-1}(t-1)e^{-(t-1)}u(t-1) + 2e^{-1}e^{-(t-1)}u(t-1).$$

The second term is just the exponential with the time shifting:

$$\mathcal{L}\{e^{-(t-1)}u(t-1)\} = \frac{e^{-s}}{s+1}.$$

The first term is $te^{-at}u(t)$ with the time shifting

$$\mathcal{L}\{(t-1)e^{-(t-1)}u(t-1)\} = \frac{e^{-s}}{(s+1)^2}.$$

Combining them and adding the constants gives

$$\mathcal{L}\{(t+1)e^{-t}u(t-1)\} = \frac{e^{-1}e^{-s}}{(s+1)^2} + \frac{2e^{-1}e^{-s}}{(s+1)}.$$

Property 7. Frequency Shifting

$$\mathcal{L}\left[f(t)e^{\lambda t}\right] = F(s-\lambda) \qquad (3.29)$$

Proof

$$\int_{0^-}^{\infty} f(t)e^{\lambda t}e^{-st}dt = \int_{0^-}^{\infty} f(t)e^{-(s-\lambda)t}dt$$
$$= F(s-\lambda).$$

Note that it looks like we took the Laplace transform using a parameter $(s-\lambda)$ instead of just s.

Property 8. Modulation

$$\mathcal{L}\left[f(t)\sin\omega_0 t\right] = \frac{1}{2j}\left[F(s-j\omega_0) - F(s+j\omega_0)\right], \quad (3.30\text{ a})$$

$$\mathcal{L}[f(t)\cos\omega_0 t] = \frac{1}{2}[F(s-j\omega_0) + F(s+j\omega_0)]. \quad (3.30\text{ b})$$

Proof

The frequency shifting theorem holds even when λ is an imaginary quantity:

$$\mathcal{L}[f(t)e^{j\omega_0 t}] = F(s-j\omega_0). \quad (3.31)$$

Using Euler's identity

$$\mathcal{L}[f(t)\sin\omega_0 t] = \mathcal{L}\left[f(t)\left(\frac{e^{j\omega_0 t} - e^{-j\omega_0 t}}{2j}\right)\right]$$

$$= \frac{1}{2j}[F(s-j\omega_0) - F(s+j\omega_0)].$$

Property 9. Time Multiplication

$$\mathcal{L}[t^n f(t)] = (-1)^n \frac{d^n F(s)}{ds^n}. \quad (3.32)$$

Proof

Start with the definition,

$$F(s) = \int_0^\infty f(t)e^{-st}dt,$$

and take the derivative with respect to s,

$$\frac{d}{ds}F(s) = \int_0^\infty -tf(t)e^{-st}dt.$$

Multiplying both sides by negative one gives:

$$(-1)^1 \frac{d}{ds}F(s) = \int_0^\infty tf(t)e^{-st}dt,$$

which proves the theorem for $n=1$. The proof for higher values of n proceeds in the same manner.

3.2.1 Stability and Laplace Transforms

We know that the Laplace domain function $F(s) = \dfrac{1}{s+2}$ will become the time-domain function $f(t) = e^{-2t} u(t)$, a function that will decay to zero in time. Now consider this function:

$$H_1(s) = \frac{1}{(s+2)(s+3)}.$$

Even before we take the inverse Laplace, we know that the time domain function will decay away as some combination of $e^{-2t} u(t)$ and $e^{-3t} u(t)$. We say that H(s) is *stable*, because we know that any reasonable input, i.e., one that is also stable, will decay away. In contrast, the function

$$H_2(s) = \frac{1}{(s+2)(s-3)}$$

is *unstable*, because it has a time domain term $e^{3t} u(t)$ that increases without end.

It does not matter that $H_2(s)$ has one stable term; if one term is unstable, the system is unstable.

Since the parameter s is complex, we can write it as $s = \sigma + j\omega$. We call the zero points in the denominator of functions like $H_1(s)$ and $H_2(s)$ the *poles*. Therefore, we can uses a graph to show the location of the H_2 poles at $s = -2$, and $s = 3$ (Fig. 3.1). Since we have seen that a pole at $s = -2$ leads to a stable result, we say the following: "All the poles must be in the left half plane for stability."

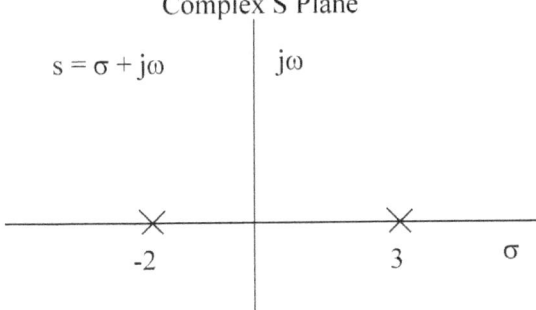

Figure 3.1. Plot of the poles of $H_2(s) = \dfrac{1}{(s+2)(s-3)}$ in the complex plane.

Property 10. Final Value Theorem

$$\lim_{t \to \infty} \{f(t)\} = \lim_{s \to 0} \{sF(s)\} \quad (3.2.10)$$

Proof

Start with the following function

$$\lim_{s \to 0} \mathcal{L}\left\{\frac{df(t)}{dt}\right\} = \lim_{s \to 0}\left[sF(s) - f(0^-)\right]$$

$$= \lim_{s \to 0}\left[sF(s)\right] - f(0^-).$$

We can also write that function as

$$\lim_{s \to 0} \mathcal{L}\left\{\frac{df(t)}{dt}\right\} = \lim_{s \to 0}\left\{\int_{0^-}^{\infty} \frac{df(t)}{dt} e^{-st} dt\right\} = \int_{0^-}^{\infty} \frac{df(t)}{dt} dt = f(\infty) - f(0^-).$$

Comparing those two equations we can conclude:

$$\lim_{s \to 0}\left[sF(s)\right] = f(\infty) = \lim_{t \to \infty} f(t).$$

This is only valid if all poles are in the left half plane.

Example 3.8 Use the final value theorem to find $\lim_{t\to\infty}\{f(t)\}$ corresponding to the following Laplace domain functions:

a. $$F_1(s) = \left[\frac{s+3}{(s+2)(s-2)}\right]$$

We cannot use the final value theorem because of the pole at $s = 2$.

b. $$F_2(s) = \left[\frac{s-3}{(s+2)(s+4)}\right]$$

$$\lim_{t\to\infty}\{f(t)\} = \lim_{s\to 0}\{sF(s)\} = \lim_{s\to 0}\left\{\frac{s(s-3)}{(s+2)(s+4)}\right\} = 0$$

c. $$F_3(s) = \left[\frac{s}{s^2+9}\right]$$

We cannot use the final value theorem because one of the poles is at $s=3$.

Example 3.9 Solve the following integral using Laplace transforms

$$x(t) = \int_0^\infty e^{-10^3 t} \sin(2\times 10^3 t)\,dt.$$

Solution

First, we rewrite the problem as

$$x(t) = \lim_{t\to\infty}\int_0^t e^{-10^3 \tau} \sin(2\times 10^3 \tau)\,d\tau.$$

Now take the Laplace transform:

$$\mathcal{L}\left\{\int_0^t e^{-10^3 \tau} \sin(2\times 10^3 \tau)\,d\tau\right\} = \frac{1}{s}\frac{2\times 10^3}{(s+10^3)^2 + (2\times 10^3)^2}.$$

Finally, we use the final value theorem,

$$\lim_{t\to\infty} x(t) = \lim_{s\to 0} sX(s) = \left.\frac{2\times 10^3}{(s+10^3)^2 + (2\times 10^3)^2}\right|_{s=0} = \frac{2}{5}\times 10^{-3}.$$

As a check, we can find the definite integral in a table:

$$\int_0^\infty e^{-10^3 t} \sin(2\times 10^3 t)\,dt = \frac{2\times 10^3}{(10^3)^2 + (2\times 10^3)^2} = \frac{2}{5}\times 10^{-3}$$

3.3 Solving Differential Equations with Laplace Transforms

Taking what we have learned so far, let us see how the Laplace transforms can be used to solve differential equations.

Example 3.10 Use Laplace transforms to solve the following differential equation with initial conditions:

$$\frac{d^2 y(t)}{dt^2} + 4\frac{dy(t)}{dt} + 3y(t) = u(t), \quad y(0) = 1,\ y'(0) = 1. \quad (3.33)$$

<u>Solution</u>

Start by taking the Laplace transform of each term. The derivative property is used to incorporate the initial conditions:

$$s^2 Y(s) - sy(0) - y'(0) + 4[sY(s) - y(0)] + 3Y(s) = \frac{1}{s}.$$

Next isolate the $Y(s)$ terms on the left:

$$Y(s)[s^2 + 4s + 3] = sy(0) + y'(0) + 4y(0) + \frac{1}{s},$$

$$= s + 5 + \frac{1}{s}.$$

We want to solve for $Y(s)$, so we write

$$Y(s) = \frac{s+5+\frac{1}{s}}{s^2+4s+3} = \frac{s+5+\frac{1}{s}}{(s+3)(s+1)} = \frac{s^2+5s+1}{s(s+3)(s+1)}.$$

Notice that we factored the denominator into individual terms. Once again, we use partial fraction expansion to break this down into terms we can look up in a table:

$$Y(s) = \frac{s^2+5s+1}{s(s+1)(s+3)} = \frac{A}{s} + \frac{B}{s+1} + \frac{C}{s+3}. \quad (3.34)$$

First, we want to isolate the terms, A, B, and C, one at a time. To isolate A, multiply through by s

$$\frac{s^2+5s+1}{(s+1)(s+3)} = A + s\frac{B}{s+1} + s\frac{C}{s+3},$$

and then evaluate at $s = 0$:

$$A = \left.\frac{s^2+5s+1}{(s+1)(s+3)}\right|_{s=0} = \frac{1}{3}.$$

Similarly, to find B, we multiply both sides of Eq. (3.34) by $s+1$, and evaluate at $s = -1$:

$$B = \left.\frac{s^2+5s+1}{s(s+3)}\right|_{s=-1} = \frac{1-5+1}{-1(2)} = \frac{-3}{-2} = \frac{3}{2}.$$

Likewise, for C we multiply by $s+3$ and evaluate at $s=-3$:

$$C = \left.\frac{s^2+5s+1}{s(s+1)}\right|_{s=-3} = \frac{9-15+1}{-3(-2)} = \frac{-5}{6}.$$

This gives us

$$Y(s) = \frac{1/3}{s} + \frac{3/2}{s+1} + \frac{-5/6}{s+3}.$$

Taking the individual terms back to the time domain gives

$$y(t) = \left(\frac{1}{3} + \frac{3}{2}e^{-t} - \frac{5}{6}e^{-3t}\right)u(t).$$

Example 3.11

Looking at the previous example, what would be different if we had

$$\frac{d^2y(t)}{dt^2} + 4\frac{dy(t)}{dt} + 3y(t) = e^{-3t}u(t) \quad y(0)=0,\ y'(0)=0\ ? \quad (3.35)$$

<u>Solution</u>

The Laplace transform of each side yields

$$Y(s)\left[s^2 + 4s + 3\right] = \frac{1}{s+3},$$

which we rewrite as

$$Y(s) = \frac{1}{(s+1)(s+3)(s+3)} = \frac{1}{(s+1)(s+3)^2}$$

Now we have a repeated root. This calls for a slightly different approach. First we expand

$$Y(s) = \frac{1}{(s+1)(s+3)^2} = \frac{A}{s+1} + \frac{B}{(s+3)^2} + \frac{C}{(s+3)}.$$

We handle the first term in the same way as we did previously. First, we cross-multiply by $(s+1)$ and then evaluate at $s=-1$ to isolate the A,

$$A = \frac{1}{(s+3)^2}\bigg|_{s=-1} = \frac{1}{(-1+3)^2} = \frac{1}{4}.$$

To get the B, we cross multiply by $(s+3)^2$,

$$\frac{1}{(s+1)} = \frac{A}{s+1}(s+3)^2 + B + \frac{C}{(s+3)}(s+3)^2,$$

and then evaluate *at* $s=-3$,

$$B = \frac{1}{(s+1)}\bigg|_{s=-3} = \frac{1}{-2}.$$

We still have no problem. Now we look for C. We start by cross-multiplying by $(s+3)$,

$$\frac{1}{(s+1)(s+3)} = \frac{A}{s+1}(s+3) + \frac{B}{(s+3)^2}(s+3) + C.$$

If we evaluate at $s=-3$, we are not going to isolate C. However, if we once again cross-multiply by $(s+3)^2$ as above

$$\frac{1}{(s+1)} = \frac{A}{s+1}(s+3)^2 + B + C(s+3)$$

and then take the derivative with respect to s

$$\frac{d}{ds}\left[\frac{1}{(s+1)}\right] = \frac{d}{ds}\left[\frac{A(s+3)^2}{s+1}\right] + C,$$

we eliminate the B. Then if we evaluate at $s=-3$, we eliminate the term containing A to give

$$C = \frac{d}{ds}\left[\frac{1}{(s+1)}\right]_{s=-3} = \left[\frac{-1}{(s+1)^2}\right]_{s=-3} = \frac{-1}{4}. \qquad (3.36)$$

Now our expansion is

$$Y(s) = \frac{1/4}{s+1} + \frac{-1/2}{(s+3)^2} + \frac{-1/4}{(s+3)},$$

and the inverse Laplace is

$$y(t) = \left(\frac{1}{4}e^{-t} - \frac{1}{2}te^{-3t} - \frac{1}{4}e^{-3t}\right)u(t).$$

Example 3.12 Find the inverse Laplace transform of:

$$F(s) = \frac{s+4}{(s+2)(s+1)^2}.$$

Assume there are no initial conditions.

Solution:

Expand $F(s)$ as

78

$$F(s) = \frac{s+4}{(s+2)(s+1)^2} = \frac{A}{s+2} + \frac{B}{(s+1)^2} + \frac{C}{(s+1)},$$

$$A = \frac{s+4}{(s+1)^2}\bigg|_{s=-2} = \frac{2}{(-1)^2} = 2,$$

$$B = \frac{s+4}{s+2}\bigg|_{s=-1} = \frac{3}{1} = 3,$$

$$C = \frac{d}{ds}\left[\frac{s+4}{s+2}\right]\bigg|_{s=-1} = \left[\frac{1(s+2)-1(s+4)}{(s+2)^2}\right]_{s=-1} = \frac{1-3}{1} = -2.$$

The above derivative was done using the formula:

$$\frac{d}{dx}\left[\frac{u(x)}{v(x)}\right] = \frac{\frac{du(x)}{dx}v(x) - u(x)\frac{dv(x)}{dx}}{v(x)^2}.$$

Check by cross-multiplying:

$$F(s) = \frac{s+4}{(s+1)(s+3)^2} = \frac{2}{s+2} + \frac{3}{(s+1)^2} + \frac{-2}{(s+1)},$$

$$s+4 = 2(s^2+2s+1) + 3(s+2) - 2(s^2+3s+2).$$

$$s^2: \quad 0 = 2-2,$$
$$s^1: \quad 1 = 4+3-6,$$
$$s^0: \quad 4 = 2+6-4.$$

The inverse of $F(s)$ is

$$f(t) = \left(2e^{-2t} + 3te^{-t} - 2e^{-t}\right)u(t).$$

Example 3.13 Solve:

$$\frac{d^2y(t)}{dt^2} + 4\frac{dy(t)}{dt} + 4y(t) = e^{-3t}u(t) \quad y(0) = 1,\ y'(0) = 0.$$

Solution

Take the Laplace transform:

$$s^2Y(s) - sy(0) + 4[sY(s) - y(0)] + 4Y(s) = \frac{1}{s+3}.$$

which gives

$$Y(s)\left[s^2 + 4s + 4\right] = s + 4 + \frac{1}{s+3},$$

$$Y(s) = \frac{s + 4 + \dfrac{1}{s+3}}{(s+2)^2}.$$

We can use different approaches to the solution.

Approach A:

$$Y(s) = \frac{(s+4)(s+3)+1}{(s+2)^2(s+3)} = \frac{s^2 + 7s + 13}{(s+2)^2(s+3)}$$

This approach is relatively straightforward. The $(s+2)^2$ term means that at some point we will have to take a derivative of a fairly complex function of s with polynomials in the numerator and denominator.

Approach B:

$$Y(s) = \frac{s + 4 + \dfrac{1}{s+3}}{(s+2)^2} = \frac{s+4}{(s+2)^2} + \frac{1}{(s+2)^2(s+3)}$$

This gives us two different terms, but they are simpler.

$$Y(s) = Y_1(s) + Y_2(s)$$

$$Y_1(s) = \frac{s+4}{(s+2)^2}, \quad Y_2(s) = \frac{1}{(s+2)^2(s+3)}.$$

First term:
Look at this very carefully before you get carried away with a lot of mathematics:

$$Y_1(s) = \frac{s+4}{(s+2)^2} = \frac{s+2}{(s+2)^2} + \frac{2}{(s+2)^2}$$
$$= \frac{1}{(s+2)} + \frac{2}{(s+2)^2}.$$
$$y_1(t) = \left(e^{-2t} + 2te^{-2t}\right)u(t).$$

Now look at the second term:
$$Y_2(s) = \frac{1}{(s+2)^2(s+3)} = \frac{C}{s+3} + \frac{D}{(s+2)^2} + \frac{E}{(s+2)}$$

$$C = \frac{1}{(s+2)^2}\bigg|_{s=-3} = \frac{1}{1^2} = 1$$

$$D = \frac{1}{(s+3)}\bigg|_{s=-2} = 1$$

$$E = \frac{d}{ds}\left[\frac{1}{(s+3)}\right]\bigg|_{s=-2} = \left[\frac{-1}{(s+3)^2}\right]\bigg|_{s=-2} = \frac{-1}{1} = -1$$

Check:
$$Y_2(s) = \frac{1}{s+3} + \frac{1}{(s+2)^2} + \frac{-1}{(s+2)} =$$
$$= \frac{(s^2+4s+4)+(s+3)-(s^2+5s+6)}{(s+3)(s+2)^2} = \frac{1}{(s+3)(s+2)^2},$$

so
$$y_2(t) = \left[e^{-3t} + te^{-2t} - e^{-2t}\right]u(t).$$

The total solution is y_1 and y_2 together:
$$y(t) = y_1(t) + y_2(t)$$
$$= \left[2te^{-2t} + e^{-2t}\right]u(t) + \left[e^{-3t} + te^{-2t} - e^{-2t}\right]u(t)$$
$$= \left[e^{-3t} + 3te^{-2t}\right]u(t).$$

3.3.1 Partial Fraction Expansion for Multiple Roots

We want to generalize the results for partial fraction expansion when we have multiple roots. If we write, for instance

$$G(s) = G_0(s) + \frac{A}{(s+\alpha)^2} + \frac{B}{(s+\alpha)}, \qquad (3.37)$$

the function $G_0(s)$ is the part that does not contain an $(s+\alpha)$ root. We solve for these by

$$A = \left[G(s)(s+\alpha)^2\right]_{s=-\alpha}$$

$$B = \frac{d}{ds}\left[G(s)(s+\alpha)^2\right]_{s=-\alpha}$$

Now consider what do to do when we have a *triple root*, as shown here:

$$G(s) = G_0(s) + \frac{A}{(s+\alpha)^3} + \frac{B}{(s+\alpha)^2} + \frac{C}{(s+\alpha)}$$

The first step is to multiply through by $(s+\alpha)^3$:

$$G(s)(s+\alpha)^3 = G_0(s)(s+\alpha)^3 + A + (s+\alpha)B + (s+\alpha)^2 C.$$

To get A, we evaluate at $s = -\alpha$

$$A = G(s)(s+\alpha)^3\Big|_{s=-\alpha}.$$

To get B, take the first derivative with respect to s

$$\frac{d}{ds}\left[G(s)(s+\alpha)^3\right]_{s=-\alpha} = G_0(s)3(s+\alpha)^2 + B + 2(s+\alpha)C,$$

and evaluate at $s = -\alpha$ to get rid of the terms we do not want:

$$B = \frac{d}{ds}\left[G(s)(s+\alpha)^3\right]_{s=-\alpha}.$$

Now, how do we get C? We take another derivative with respect to s

$$\frac{d^2}{ds^2}\left[G(s)(s+\alpha)^3\right]_{s=-\alpha} = G_0(s)3\cdot 2(s+\alpha) + 2C,$$

and evaluate at $\alpha = -3$ to isolate the C parameter. But notice the difference.

$$C = \frac{1}{2}\frac{d^2}{ds^2}\left[G(s)(s+\alpha)^3\right]_{s=-\alpha}.$$

C has acquired a factor ½. Guess what D would be if we increase by one power:

$$G(s) = G_0(s) + \frac{A}{(s+\alpha)^4} + \frac{B}{(s+\alpha)^3} + \frac{C}{(s+\alpha)^2} + \frac{D}{(s+\alpha)}.$$

By taking the above argument one step further, we can conclude.

$$D = \frac{1}{3\cdot 2}\frac{d^3}{ds^3}\left[G(s)(s+\alpha)^4\right]_{s=-\alpha}.$$

Example 3.14 In the following partial fraction expansion, solve for B:

$$H(s) = \frac{1}{s(s+1)^5} = \frac{A}{s} + \frac{B}{(s+1)} + \frac{C}{(s+1)^2} + \frac{D}{(s+1)^3} + \frac{E}{(s+1)^4} + \frac{F}{(s+1)^5}$$

Solution

For this problem, $N = 5$.

$$B = \frac{1}{(5-1)!}\left(\frac{d}{ds}\right)^{5-1}\left[H(s)(s+1)^5\right]_{s=-1} = \frac{1}{4!}\left(\frac{d}{ds}\right)^4\left[\frac{1}{s}\right]_{s=-1}$$

$$= \frac{1}{4\cdot 3\cdot 2}\left(\frac{d}{ds}\right)^4\left[\frac{-1\cdot-2\cdot-3\cdot-4}{s^5}\right]_{s=-1} = \left[\frac{1}{s^5}\right]_{s=-1} = -1$$

Example 3.15 Solve for C:

$$F(s) = \frac{(s+2)}{(s+3)(s+1)^2} = \frac{A}{(s+3)} + \frac{B}{(s+1)^2} + \frac{C}{(s+1)}$$

Solution

$$C = \frac{d}{ds}\left[\frac{(s+2)}{(s+3)}\right]_{s=-1} = \left[\frac{1(s+2)-1(s+3)}{(s+3)^2}\right]_{s=-1}$$

$$= \left[\frac{-1}{(s+3)^2}\right]_{s=-1} = -\frac{1}{4}.$$

Example 3.16 A system is described by the following equation:

$$\frac{d^2 y(t)}{dt^2} + 5\frac{dy(t)}{dt} + 4y(t) = f(t).$$

Find the impulse response.

Solution

In finding the impulse response, initial conditions play no role. We start by taking the Laplace transform:

$$s^2 Y(s) + 5sY(s) + 4Y(s) = 1.$$

$$Y(s) = \frac{1}{s^2 + 5s + 4} = \frac{A}{(s+1)} + \frac{B}{(s+4)}$$

$$= \frac{1/3}{(s+1)} - \frac{1/3}{(s+4)}.$$

So

$$h(t) = \frac{1}{3}\left(e^{-t} - e^{-4t}\right)u(t)$$

3.4 Inverse of Laplace Transforms with Complex Roots

So far, we have been dealing with the inverse of real roots. We also need a method to find the inverse of pairs of complex conjugate roots.

3.4.1 Inverse of a Complex Conjugate pair.

We start with an example to illustrate the inverse of a complex conjugate pair.

Example 3.17. Use Laplace transforms to solve the following differential equation with no initial conditions.

$$\frac{d^2 y(t)}{dt^2} + 2\frac{dy(t)}{dt} + 4y(t) = e^{-2t}u(t) \quad y(0) = 0, \ y'(0) = 0$$

Solution

The Laplace transform is,

$$s^2 Y(s) + 2sY(s) + 4Y(s) = \frac{1}{s+2},$$

or

$$Y(s) = \frac{1}{(s^2 + 2s + 4)(s+2)}. \tag{3.38}$$

There is second order term. When we try to factor that term, we get

$$p_{1,2} = \frac{-2 \pm \sqrt{2^2 - 4 \cdot 4}}{2} = \frac{-2 \pm 2j\sqrt{1-4}}{2} = -1 \pm j\sqrt{3},$$

which is a complex conjugate pair:

$$p = -1 + j\sqrt{3} \quad p^* = -1 - j\sqrt{3}. \tag{3.39}$$

Nothing says we cannot use the same partial fraction expansion for complex numbers, so rewrite Eq. (3.4.2) as

$$Y(s) = \frac{1}{(s^2 + 2s + 4)(s+2)} = \frac{A}{s+2} + \frac{k}{s-p} + \frac{k^*}{s-p^*}. \tag{3.40}$$

Notice that we are assuming the coefficient k^* is the complex conjugate of the coefficient k. We will see that this proves to be true. Solving for A,

$$A = \frac{1}{(s^2+2s+4)}\bigg|_{s=-2} = \frac{1}{4-4+4} = \frac{1}{4}.$$

We solve for k using the same basic procedure even though we have complex roots:

$$k = \frac{1}{s+2}\frac{1}{s-p^*}\bigg|_{s=p} = \frac{1}{s+2}\frac{1}{s-(-1-j\sqrt{3})}\bigg|_{s=(-1+j\sqrt{3})} \quad (3.41)$$

$$= \frac{1}{(-1+j\sqrt{3})+2}\frac{1}{2j\sqrt{3}} = \frac{1}{2\angle 60°}\frac{1}{2\sqrt{3}\angle 90°} = \frac{1}{4\sqrt{3}}\angle -150°.$$

Now solve for k^*

$$k^* = \frac{1}{s+2}\frac{1}{s-p}\bigg|_{s=p^*} = \frac{1}{s+2}\frac{1}{s-(-1+j\sqrt{3})}\bigg|_{s=(-1-j\sqrt{3})}$$

$$= \frac{1}{(-1-j\sqrt{3})+2}\frac{1}{-2j\sqrt{3}} = \frac{1}{2\angle -60°}\frac{1}{2\sqrt{3}\angle -90°} = \frac{1}{4\sqrt{3}}\angle 150°.$$

Notice that we did not have to explicitly solve for k^*, because it is always the complex conjugate of k. Taking Eq. (3.4.4) to the time domain gives:

$$y(t) = \left[\frac{1}{4}e^{-2t} + \frac{e^{-j150°}}{4\sqrt{3}}e^{(-1+j\sqrt{3})} + \frac{e^{j150°}}{4\sqrt{3}}e^{(-1-j\sqrt{3})}\right]$$

$$= \left[\frac{1}{4}e^{-2t} + \frac{e^{-t}}{4\sqrt{3}}\left(e^{j(\sqrt{3}t-150°)} + e^{-j(\sqrt{3}t-150°)}\right)\right] \quad (3.42)$$

$$= \left[\frac{1}{4}e^{-2t} + \frac{1}{2\sqrt{3}}e^{-t}\cos(\sqrt{3}t-150°)\right]u(t).$$

3.4.2 The Cosine Method

We probably want to look for a more systematic way to determine the inverse Laplace of the complex conjugate pair. If we take the complex root with the positive imaginary value to be
$$p = \alpha + j\omega,$$
we can generalize the above procedure by writing
$$y(t) = \left[\frac{1}{4}e^{-2t} + 2|k|e^{\alpha t}\cos(\omega t + \angle k)\right]u(t).$$

For this problem
$$\alpha = -1; \quad \omega = \sqrt{3},$$
$$|k| = \frac{1}{4\sqrt{3}}, \quad \angle k = 150^\circ.$$

In general, if a Laplace domain function $F(s)$ contains a complex conjugate pair, the following procedure can be used to find the time-domain term due to the complex pair. The function $F(s)$ is written

$$F(s) = \frac{F_0(s)}{(s+p)(s+p^*)}, \qquad (3.43)$$

where $F_0(s)$ is everything but the complex pair.

The *cosine method* consists of the following steps:
a. Write the roots as
$$p = \alpha + j\omega$$
(Remember: p always contains the positive imaginary part.)
b. Calculate
$$k = \left.\frac{F_0(s)}{(s-p^*)}\right|_{s=p} = |k|\angle k$$
c. The part due to the complex roots is
$$y_c(t) = 2|k|e^{\alpha t}\cos(\omega t + \angle k)u(t).$$

Example 3.18. Redo Example 3.4.1 using the cosine method.

<u>Solution</u>

a. $p = -1 + j\sqrt{3}$, so $\alpha = -1$ and $\omega = j\sqrt{3}$.

b. $k = \dfrac{1}{s+2} \dfrac{1}{s-p^*} \bigg|_{s=p} = \dfrac{1}{4\sqrt{3}} \angle -150°$,

so $|k| = \dfrac{1}{4\sqrt{3}}$ and $\angle k = \angle -150°$.

c. $y_c(t) = 2|k|e^{\alpha t}\cos(\omega t + \angle k)u(t)$.

3.4.3. The Sine Method

There might be an even easier way to solve a problem like Example 3.4.1. Look at the calculation of k in Eq. (3.4.5):

$$k = \dfrac{1}{s+2} \dfrac{1}{s-\left(-1-j\sqrt{3}\right)}\bigg|_{s=\left(-1+j\sqrt{3}\right)} = \dfrac{1}{\left(-1+j\sqrt{3}\right)+2} \cdot \dfrac{1}{2j\sqrt{3}}.$$

Notice that the term on the far right could be written:

$$\dfrac{1}{2j\sqrt{3}} = \dfrac{1}{2j\omega}, \quad (3.4.8)$$

because

$$\dfrac{1}{s-p^*}\bigg|_{s=p} = \dfrac{1}{p-p^*} = \dfrac{1}{2j\,\text{Im}[p]} = \dfrac{1}{2j\omega}.$$

There will always be a purely imaginary term like this, because whenever we solve for the coefficient of the $s - p$ root, the $s - p^*$ will be in the denominator. If we know that, we can solve for the simpler term

$$g = \frac{1}{s+2}\Big|_{s=p} = \frac{1}{(1+j\sqrt{3})} = \frac{1}{2}\angle -60°. \tag{3.44}$$

In relation to our previous way of solving for the coefficients:

$$k = \frac{g}{2j\omega}, \quad k^* = \frac{g^*}{-2j\omega} \tag{3.4.10}$$

So now, Eq. (3.4.4) becomes

$$Y(s) = \frac{1/4}{s+2} + \frac{1}{2j\omega}\left[\frac{g}{s-p} + \frac{g^*}{s-p^*}\right],$$

which in the time domain is:

$$y(t) = \frac{1}{4}e^{-2t}u(t) + \frac{|g|}{\omega}\left[\frac{e^{\angle g}e^{(\alpha+j\omega)t} - e^{\angle g}e^{(\alpha-j\omega)t}}{2j}\right]u(t)$$

$$= \frac{1}{4}e^{-2t}u(t) + \frac{|g|}{\omega}e^{\alpha t}\sin(\omega t + \angle g)u(t) \tag{3.45}$$

$$= \frac{1}{4}e^{-2t}u(t) + \frac{1}{2\sqrt{3}}e^{-t}\sin(\sqrt{3}t - 60°)u(t).$$

Notice that since

$$\sin(\theta \pm 90°) = \pm\cos(\theta°),$$

we could write:

$$\sin(\sqrt{3}t - 150° + 90°) = \cos(\sqrt{3}t - 150°).$$

The *sine method* consists of the following steps:

a. Write the roots as

$$p = \alpha + j\omega \quad p^* = \alpha - j\omega$$

As before, p always contains the positive imaginary part.

b. Calculate

$$g = F'(s)\big|_{s=p} = |g|\angle g$$

c. The part due to the complex roots is

$$y_s(t) = \frac{|g|}{\omega} e^{\alpha t} \sin(\omega t + \angle g) u(t)$$

Example. 3.19. Find the inverse Laplace transform of the following function:
$$F(s) = \frac{s+2}{s^2 + 3s + 5}.$$

<u>Solution (Sine method):</u>

a. First find the root
$$p = \frac{-3 + \sqrt{3^2 - 4(5)}}{2} = \frac{-3 + j\sqrt{11}}{2} = -1.5 + j1.66.$$

b. Now determine the complex coefficient g
$$g = s + 2 \big|_{s = -1.5 + j1.66} = 0.5 + j1.66 = 1.73 \angle 73°.$$

c. The corresponding time-domain term is
$$f(t) = \frac{1.73}{1.66} e^{-1.5t} \sin(1.66t + 73°) u(t)$$
$$= 1.04 e^{-1.5t} \sin(1.66t + 73°) u(t).$$

Example 3.20. Find the inverse Laplace transform of
$$X(s) = \frac{5s + 13}{s(s^2 + 4s + 13)}$$

<u>Solution (Sine method)</u>

$$X(s) = \frac{5s + 13}{s(s^2 + 4s + 13)} = \frac{A}{s} + \frac{M}{(s^2 + 4s + 13)}.$$

Notice that the M is merely a place holder. We will not explicitly solve for M, but instead use the sine method.

$$A = \frac{5s + 13}{(s^2 + 4s + 13)} \bigg|_{s=0} = \frac{13}{13} = 1$$

90

a. $p = \dfrac{-4+\sqrt{4^2-4\cdot 13}}{2} = \dfrac{-4+2\sqrt{4-13}}{2}$

$= -2+j3$

b. $g = \dfrac{5s+13}{s}\bigg|_{s=p} = \dfrac{3+j15}{(-2+j3)} = \dfrac{15.3\angle 79°}{3.6\angle 124°} = 4.25\angle -45°$.

c. $x_s(t) = \dfrac{4.25}{3} e^{-2t}\sin(3t-45°)u(t),$

The total solution is,

$$x(t) = \left[1 + 1.41e^{-2t}\sin(3t-45°)\right]u(t).$$

Example 3.21. Find the zero-state response of

$$y'' + 3y' + 2y = \sin(t)u(t). \qquad (3.46)$$

Zero-state means no initial conditions.

<u>Solution:</u>

Taking the Laplace transform gives

$$Y(s)(s^2 + 3s + 2) = \dfrac{1}{s^2+1}$$

$$Y(s) = \dfrac{1}{(s^2+3s+2)(s^2+1)}$$

$$= \dfrac{A}{(s+1)} + \dfrac{B}{(s+2)} + \dfrac{M}{(s^2+1)}$$

$$A = \dfrac{1}{(s+2)(s^2+1)}\bigg|_{s=-1} = \dfrac{1}{1\cdot 2} = \dfrac{1}{2}$$

$$B = \dfrac{1}{(s+1)(s^2+1)}\bigg|_{s=-2} = \dfrac{1}{-1\cdot 5} = -\dfrac{1}{5}$$

Using the sine method, we have

a. $p = j \Rightarrow \alpha = 0, \omega = 1$

b. $g = \dfrac{1}{(s+1)(s+2)}\bigg|_{s=i} = \dfrac{1}{(i+1)(i+2)}$

$= \dfrac{1}{\sqrt{2}\angle 45^\circ \cdot \sqrt{5}\angle 27^\circ} = \dfrac{1}{\sqrt{10}} \angle -72^\circ$

c. $y_s(t) = \dfrac{1/\sqrt{10}}{1} e^{-0t} \sin(t - 72^\circ) u(t)$

So

$$y(t) = \left[\dfrac{1}{2}e^{-t} - \dfrac{1}{5}e^{-2t} + 0.316 \sin(t - 72^\circ)\right] u(t)$$

Example 3.22. What would be different if the input of the previous example was

$$f(t) = \sin(t-2)u(t-2)?$$

Solution

From time-invariance, we can just use the above example and delay by two seconds,

$$y(t) = \left[\dfrac{1}{2}e^{-(t-2)} - \dfrac{1}{5}e^{-2(t-2)} + 0.316 \sin((t-2) - 72^\circ)\right] u(t-2)$$

Example 3.23. What would be different if the input of the previous example were

$$f(t) = \sin(t)u(t-2)?$$

Solution

We begin by writing

$$f(t) = \sin(t)u(t-2) = \sin((t-2)+2)u(t-2).$$

Be careful, the two in the $\sin\left[(t-2)+2\right]$ is in radians!

$$2 \text{ radians} \times \frac{180^0}{\pi \text{ radians}} = 114^0.$$

So actually, we could write
$$f(t) = \sin(t - 2 + 114^\circ)u(t-2);$$
then we would solve the problem using
$$f'(t) = \sin(t + 114^\circ)u(t)$$
$$= \sin(t)\cos(114^\circ)u(t) + \cos(t)\sin(114^\circ)u(t)$$
$$= -0.4\sin(t)u(t) + 0.91\cos(t)u(t)$$
as our input, and then shift the answer by two seconds when we finish. Notice that the 114° on the input cannot just be eliminated and added at the end, as was the case when the input was a non-causal sinusoid and we used phasors.

Example 3.24. Solve $y'' + 2y' + 5y = u(t)\quad y(0) = 1, y'(0) = 1$

Solution:

The Laplace transform is

$$s^2Y(s) - sy(0) - y'(0) + 2(Y(s) - y(0)) + 5Y(s) = \frac{1}{s}$$

$$Y(s)(s^2 + 2s + 5) = s + 3 + \frac{1}{s}$$

$$Y(s) = \frac{s^2 + 3s + 1}{s(s^2 + 2s + 5)} = \frac{A}{s} + \frac{M}{(s^2 + 2s + 5)}$$

$$A = \frac{1}{5}$$

Using the sine method, we have:

a. $\quad p = \dfrac{-2 + \sqrt{2^2 - 4 \cdot 5}}{2} = -1 + j2$

b.
$$g = \left.\frac{s^2+3s+1}{s}\right|_{s=-1+j2} = \frac{(1-4-j4)+(-3+j6)+1}{-1+j2}$$

$$= \frac{-5+j2}{-1+j2} = \frac{5.39\angle 158°}{2.24\angle 117°} = 2.41\angle 41°$$

c.
$$y_s(t) = \frac{2.41}{2}e^{-t}\sin(2t+41°)u(t).$$

The total response is
$$y(t) = \left[0.2+1.2e^{-t}\sin(2t+41°)\right]u(t).$$

Example 3.25. Find the inverse Laplace transform of
$$F(s) = \frac{10^3}{s(s^2+5\times 10^3 s+1.2\times 10^7)}$$

<u>Solution</u>

$$F(s) = \frac{10^3}{s(s^2+5\times 10^3+1.2\times 10^7)} = \frac{A}{s} + \frac{M}{(s^2+5\times 10^3+1.2\times 10^7)}$$

$$A = \left.\frac{10^3}{(s^2+5\times 10^3+1.2\times 10^7)}\right|_{s=0} = 0.83\times 10^{-4}$$

a. $p = \left(\dfrac{-5+\sqrt{5^2-4\cdot 12}}{2}\times 10^3\right) = (-2.5+j2.4)\times 10^3$

b. $g = \left.\dfrac{10^3}{s}\right|_{s=p} = \dfrac{10^3}{(-2.5+j2.4)\times 10^3}$

$$= \frac{1}{3.47\angle 136°} = 0.288\angle -136°$$

c. $f_s(t) = \dfrac{0.288}{2.4\times 10^3} e^{-2.5\times 10^3 t}\sin(2.4\times 10^3 t - 136°)u(t)$

$$f(t) = \left[0.83\times 10^{-4} + 1.2\times 10^{-4} e^{-2.5\times 10^3 t}\sin(2.4\times 10^3 t - 136°)\right]u(t)$$

Example 3.26. Solve for y(t).

$$\frac{d^2y(t)}{dt^2}+6\frac{dy(t)}{dt}+8y(t)=\frac{df(t)}{dt}+5f(t),$$

$$y(0)=0,\ y'(0)=0,\quad f(t)=e^{-3t}u(t)$$

Solution

First take the equation to the Laplace domain.

$$Y(s)(s^2+6s+8)=(s+5)F(s)=(s+5)\frac{1}{s+3}. \quad (3.47)$$

Notice that we first evaluated

$$\mathcal{L}\left\{\frac{df(t)}{dt}+5f(t)\right\}=(sF(s)+5F(s))=(s+5)F(s).$$ When dealing with the input function, there are no initial conditions! Then we add

$$\mathcal{L}\{f(t)\}=\frac{1}{s+3}$$ to get the left side of Eq. (3.4.12). Now we continue as usual.

$$Y(s)=\frac{s+5}{(s+2)(s+4)(s+3)}=\frac{A}{s+2}+\frac{B}{s+3}+\frac{C}{s+4}$$

$$A=\frac{s+5}{(s+3)(s+4)}\bigg|_{s=-2}=\frac{3}{(1)(2)}=\frac{3}{2};$$

$$B=\frac{s+5}{(s+2)(s+4)}\bigg|_{s=-3}=\frac{2}{(-1)(1)}=-2$$

$$C=\frac{s+5}{(s+2)(s+3)}\bigg|_{s=-4}=\frac{1}{(-2)(-1)}=2$$

Going to the time domain,

$$y(t)=\left(1.5e^{-2t}-2e^{-3t}+2e^{-4t}\right)u(t)$$

Note:

$$\mathcal{L}\left\{\frac{d}{dt}e^{-3t}u(t)+5e^{-3t}u(t)\right\} = \mathcal{L}\left\{-3e^{-3t}u(t)+\delta(t)+5e^{-3t}u(t)\right\}$$

$$= 1 - \frac{3}{s+3} + \frac{5}{s+3} = \frac{s+5}{s+3}$$

3.5 Block Diagrams

Suppose we have a system described by the differential equation

$$\frac{d^2 y(t)}{dt^2} + 5\frac{dy(t)}{dt} + 2y(t) = \frac{df(t)}{dt} + 3f(t), \qquad (3.48)$$

$$y(0) = 0,\ y'(0) = 0$$

We can take the Laplace transform to get

$$s^2 Y(s) + 5sY(s) + 2Y(s) = sF(s) + 3F(s).$$

From this, we can determine the *transfer function*

$$H(s) = \frac{Y(s)}{F(s)} = \frac{s+3}{s^2 + 5s + 2}.$$

We can take the inverse Laplace of the transfer function to get the impulse response

$$h(t) = \mathcal{L}^{-1}\{H(s)\}.$$

It will prove useful to make the following diagram of this system:

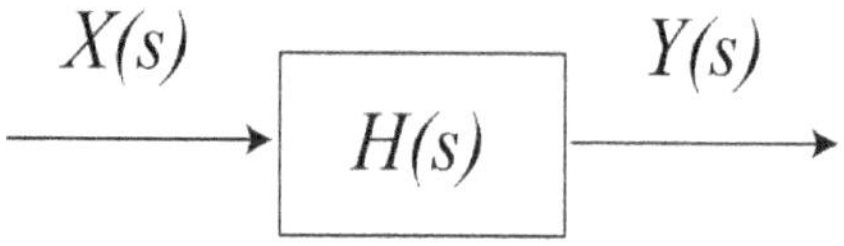

$$X(s) \longrightarrow \boxed{H(s)} \longrightarrow Y(s)$$

Figure 3.2. Block diagram of a system described by the transfer function $H(s)$.

This is referred to as a *block diagram*. Block diagrams are extremely useful when creating systems from sub-systems.

3.5.1. Block Diagram Analysis.

The following that result from the fact that we are dealing with linear systems are useful.

Property A. Series Connection

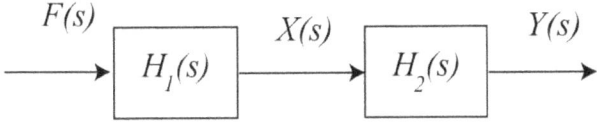

Figure 3.3. Series connection of two systems.

If the output of the first block is $X(s)$, then we say
$$X(s) = H_1(s)F(s).$$
The output of the second block is
$$Y(s) = H_2(s)X(s)$$
$$= H_2(s)\left[H_1(s)F(s)\right].$$
Therefore, for the system in Fig. 3.5.2, the total transfer function of the series of blocks is
$$H_S(s) = \frac{Y(s)}{F(s)} = H_1(s)H_2(s). \tag{3.49}$$

Property B. Parallel Connection

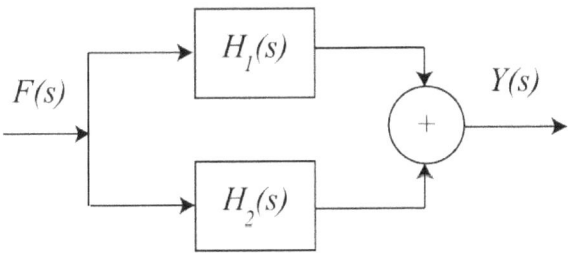

Figure 3.4 Parallel connections of two systems.

For the parallel connection shown in Fig. 3.4, we see that the output is
$$Y(s) = H_1(s)F(s) + H_2(s)F(s)$$
$$= (H_1(s) + H_2(s))F(s).$$
Therefore, the total transfer function of the parallel blocks is given by
$$H_P(s) = \frac{Y(s)}{F(s)} = H_1(s) + H_2(s) \quad (3.50)$$

Property C. Negative Feedback Loop

The block diagram of Fig. 3.5 contains *negative feedback*.

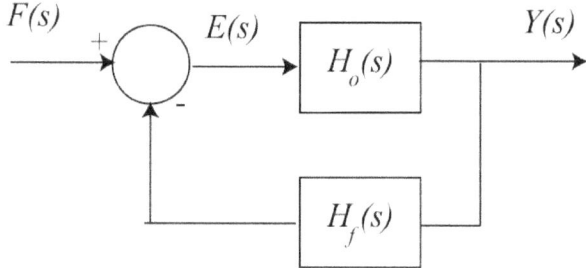

Figure 3.5. A negative feedback system

The total transfer function of the simple negative feedback circuit can be calculated easily. The system consists of the *open-loop* transfer function, $H_o(s)$ and the *feedback* transfer function $H_f(s)$. In determining the total transfer function of Fig. 3.5.4, it is expedient to define an *error function*, $E(s)$, right after the summation. We solve by noticing
$$E(s) = F(s) - H_f(s)Y(s).$$
$$Y(s) = H_0(s)E(s)$$
$$= H_0(s)[F(s) - H_f(s)Y(s)].$$
We then group together the *Y(s)* terms
$$Y(s) + H_0(s)H_f(s)Y(s) = H_0(s)F(s)$$

$$Y(s) = \frac{H_0(s)}{1+H_f(s)H_0(s)} F(s).$$

$$H_T(s) = \frac{Y(s)}{F(s)} = \frac{H_0(s)}{1+H_f(s)H_0(s)}. \tag{3.51}$$

This property is so universal, it is worth memorizing.

Example 3.27. Refer to the block diagram in Fig. 3.5.

a. What is the total transfer function if

$$H_0(s) = \frac{1}{s+3}, \quad H_f(s) = \frac{2}{s}.$$

b). What is its impulse response?
c). What is the step response?

Solution

a. $$H_T(s) = \frac{\left(\frac{1}{s+3}\right)}{1+\left(\frac{2}{s}\right)\left(\frac{1}{s+3}\right)} = \frac{s}{s(s+3)+2} = \frac{s}{s^2+3s+2}$$

b. $$H_T(s) = \frac{s}{s^2+3s+2} = \frac{A}{s+1} + \frac{B}{s+2}$$

$$A = \left.\frac{s}{s+2}\right|_{s=-1} = \frac{-1}{1}, \quad B = \left.\frac{s}{s+1}\right|_{s=-2} = \frac{-2}{-1} = 2$$

$$h(t) = \left(-e^{-t} + 2e^{-2t}\right)u(t)$$

c.

$$Y_{step}(s) = H_T(s)\frac{1}{s}$$

$$= \frac{s}{s^2+3s+2}\cdot\frac{1}{s} = \frac{1}{s^2+3s+2} = \frac{A}{s+1} + \frac{B}{s+2} + \frac{C}{s}$$

$$A = \frac{1}{s}\frac{1}{(s+2)}\bigg|_{s=-1} = -1, \quad B = \frac{1}{s(s+1)}\bigg|_{s=-2} = \frac{1}{3},$$

$$C = \frac{1}{(s+1)(s+2)}\bigg|_{s=-} = \frac{1}{2}.$$

$$y_{step}(t) = \left(\frac{1}{2} - e^{-t} + \frac{1}{3}e^{-2t}\right)u(t)$$

Example 3.28. Find the total transfer function corresponding to the following block diagrams.

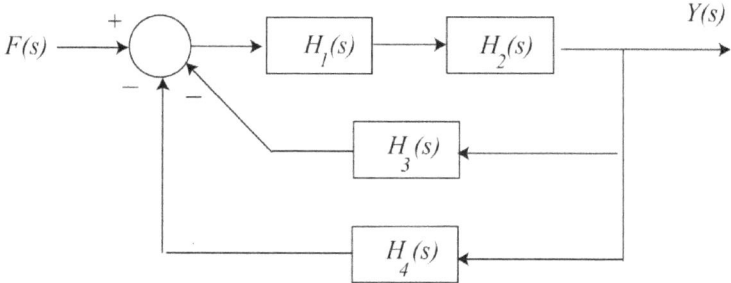

Figure 3.6. Block diagram 1

Solution

From Property A, we can multiply together the two blocks
$$H_o(s) = H_1(s)H_2(s).$$
From Property B, we can add together the two feedback loops
$$H_f(s) = H_3(s) + H_4(s).$$
From Property C, we can write the total transfer function,
$$H_T(s) = \frac{(H_1 H_2)}{1 + (H_1 H_2)(H_3 + H_4)}.$$

Example 3.29. Find the total transfer function corresponding to the following block diagrams.

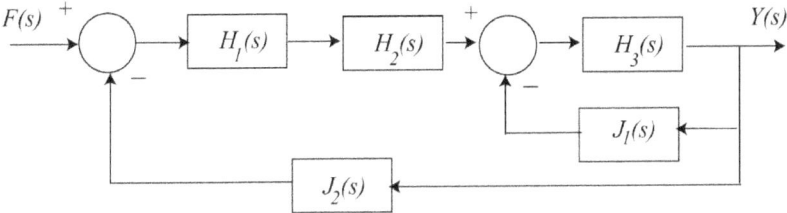

Figure 3.7. Block diagram 2.

Write the small loop on the right as
$$H' = \frac{H_3}{1+H_3J_1}.$$

Now we have
$$H_0 = \frac{H_1H_2H_3}{1+H_3J_1}.$$

So
$$H_T = \frac{\dfrac{H_1H_2H_3}{1+H_3J_1}}{1+\dfrac{H_1H_2H_3}{1+H_3J_1}(J_2)}$$
$$= \frac{H_1H_2H_3}{1++H_3J_1+H_1H_2H_3J_2}.$$

Example 3.30. Find the impulse and step responses of the following system,

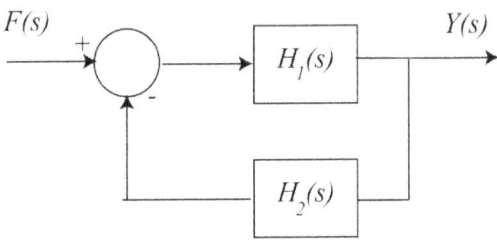

Figure 3.8. System 1.

101

for $H_0(s) = H_1(s) = \dfrac{1}{s+2}$, $H_f(s) = H_2(s) = \dfrac{1}{s+3}$.

Solution:

$$H_T(s) = \dfrac{H_0}{1+H_0 H_f} = \dfrac{\dfrac{1}{s+2}}{1+\dfrac{1}{s+2}\dfrac{1}{s+3}}$$

$$= \dfrac{s+3}{(s+2)(s+3)+1} = \dfrac{s+3}{s^2+5s+7}$$

Using the sine method, we get:

a. $p = \dfrac{-5+\sqrt{5^2 - 4\cdot 7}}{2} = -\dfrac{5}{2} + j\dfrac{\sqrt{3}}{2}$.

b. $g = s+3\big|_{s=-\frac{5}{2}+j\frac{\sqrt{3}}{2}} = -\dfrac{5}{2} + j\dfrac{\sqrt{3}}{2} + 3$

$= \dfrac{1}{2} + j\dfrac{\sqrt{3}}{2} = 1\angle 60°$

c. $h(t) = \dfrac{1}{\sqrt{3}/2} e^{-\frac{5}{2}t} \sin\left(\dfrac{\sqrt{3}}{2}t + 60°\right)$

$= 1.154 e^{-\frac{5}{2}t} \sin\left(\dfrac{\sqrt{3}}{2}t + 60°\right)$.

The step response is

$$Y_s(s) = \dfrac{s+3}{s^2+5s+7}\dfrac{1}{s} = \dfrac{A}{s} + \dfrac{M}{s^2+5s+7}$$

$$A = \dfrac{3}{7} = 0.428$$

$$g = \left.\frac{s+3}{s}\right|_{s=-\frac{5}{2}+j\frac{\sqrt{3}}{2}} = \frac{-\frac{5}{2}+j\frac{\sqrt{3}}{2}+3}{-\frac{5}{2}+j\frac{\sqrt{3}}{2}} = \frac{1+j\sqrt{3}}{-5+j\sqrt{3}}$$

$$= \frac{2\angle 60^\circ}{5.29\angle 19^\circ} = 0.378\angle 41^\circ$$

$$y(t) = 0.428u(t) + \frac{3.78}{\sqrt{3}} e^{-\frac{5}{2}t} \sin\left(\frac{\sqrt{3}}{2}t + 41^\circ\right) u(t)$$

$$= 0.428u(t) + 2.18 e^{-\frac{5}{2}t} \sin\left(\frac{\sqrt{3}}{2}t + 41^\circ\right) u(t)$$

3.5.2 Block Diagram Analysis Using Error Functions

We have introduced some properties that can be used to analyze block diagrams. However, there are times when this is impractical. Here is an alternative approach:

a. Define an auxiliary variable after each adder.
b. Write an algebraic expression for each adder.
c. Solve to get output as a function only of input.

Example 3.31. Find the total transfer function of the following block diagram utilizing the intermediate parameters E_1 and E_2.

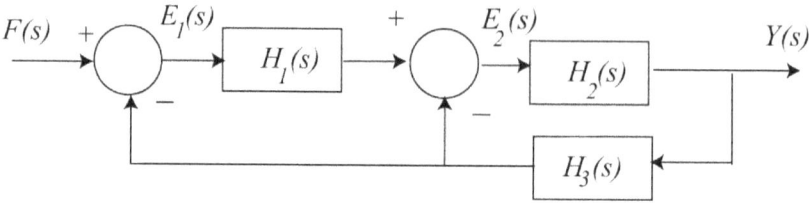

Figure 3.9. Block diagram

Solution

$$E_1 = F - H_3Y$$
$$E_2 = H_1E_1 - H_3Y$$
$$Y = H_2E_2$$
$$= H_2(H_1E_1 - H_3Y)$$
$$= H_2H_1(F - H_3Y) - H_2H_3Y$$
$$Y(1 + H_2H_3Y + H_1H_2H_3Y) = H_1H_2F$$

$$H_T = \frac{Y}{F} = \frac{H_1H_2}{1 + H_2H_3 + H_1H_2H_3}$$

Alternative Solution Using the Theorems

First of all, we to redraw the diagram

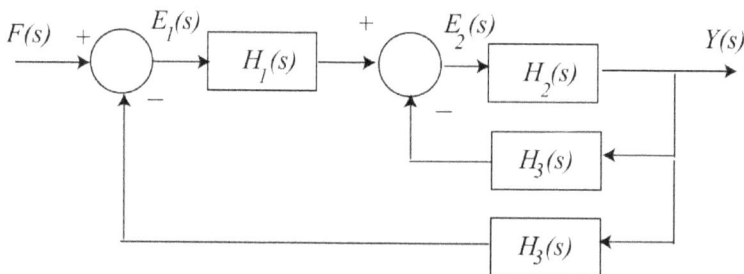

Figure 3.9a. Redraw of Fig. 3.9.

Then we can evaluate the loop in the upper right corner

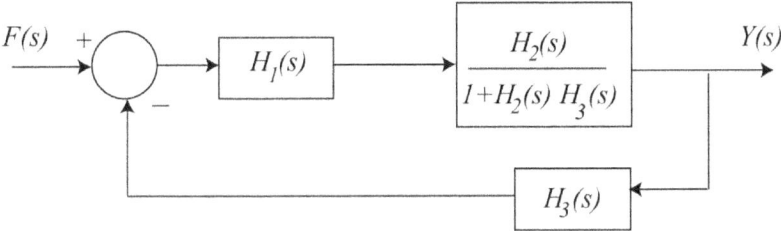

Figure 3.9b. Redraw of Fig. 3.9a

Finally,

$$H_T = \frac{\left(\dfrac{H_1 H_2}{1+H_2 H_3}\right)}{1+H_3\left(\dfrac{H_1 H_2}{1+H_2 H_3}\right)} = \frac{H_1 H_2}{1+H_2 H_3 + H_1 H_2 H_3}$$

Example 3.32. Find the transfer function of the following block diagram

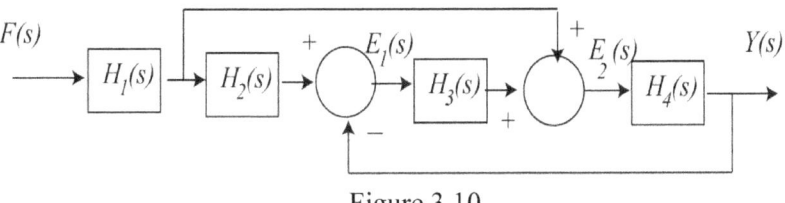

Figure 3.10.

Solution

To solve this, it would be hard to apply the theorems.
$$E_1 = H_1 H_2 F - Y$$
$$E_2 = H_1 F + H_3 E_1$$
$$= H_1 F + H_3 \left(H_1 H_2 F - Y\right)$$

105

$$Y = H_4 E_2$$
$$= H_4 \left(H_1 F + H_3 \left(H_1 H_2 F - Y \right) \right)$$
$$= \left(H_1 H_4 + H_1 H_2 H_3 H_4 \right) F - H_3 H_4 Y$$
$$Y \left(1 + H_3 H_4 \right) = \left(H_1 H_4 + H_1 H_2 H_3 H_4 \right) F$$
$$H_T = \frac{Y}{F} = \frac{H_1 H_4 + H_1 H_2 H_3 H_4}{1 + H_3 H_4}$$

Table 3.1 Some Common Laplace Transform Pairs

$f(t)$	$F(s)$
$\delta(t)$	1
$u(t)$	$\dfrac{1}{s}$
$t^n u(t)$	$\dfrac{n!}{s^{n+1}}$
$e^{-\alpha t} u(t)$	$\dfrac{1}{s+\alpha}$
$t^n e^{-\alpha t} u(t)$	$\dfrac{n!}{(s+\alpha)^{n+1}}$
$\sin(\omega t) u(t)$	$\dfrac{\omega}{s^2+\omega^2}$
$\cos(\omega t) u(t)$	$\dfrac{s}{s^2+\omega^2}$
$e^{-\alpha t}\sin(\omega t) u(t)$	$\dfrac{\omega}{(s+\alpha)^2+\omega^2}$
$e^{-\alpha t}\cos(\omega t) u(t)$	$\dfrac{s+\alpha}{(s+\alpha)^2+\omega^2}$

Table 3.2 Some Properties of Laplace Transforms.

$f(t)$	$F(s)$
$f(t-t_0)$	$e^{-st_0}F(s)$
$f(t)e^{\lambda t}$	$F(s-\lambda)$
$t^n f(t)$	$(-1)^n \dfrac{d^n}{ds^n}F(s)$
$f(at)$	$\dfrac{1}{a}F\left(\dfrac{s}{a}\right)$
$f(t)e^{\lambda t}$	$F(s-\lambda)$
$f(t)\cos(\omega_0 t)$	$\dfrac{1}{2}\left[F(s+j\omega_0)+F(s-j\omega_0)\right]$
$f(t)\sin(\omega_0 t)$	$\dfrac{j}{2}\left[F(s+j\omega_0)-F(s-j\omega_0)\right]$
$\dfrac{d}{dt}f(t)$	$sF(s)-f(0^-)$
$\dfrac{d^2}{dt^2}f(t)$	$s^2F(s)-sf(0^-)-f^{(1)}(0^-)$
$f_1(t)*f_2(t)$	$F_1(s)F_2(s)$
$\int_0^t f(\tau)d\tau$	$\dfrac{1}{s}F(s)$
$\lim\limits_{t\to\infty}\{f(t)\}$	$\lim\limits_{s\to 0}\{sf(t)\}$

References

1. Z. Gajic, *Linear Dynamic Systems and Signals,* Upper Saddle River, NJ: Prentice Hall, 2003.

2. P. D. Cha and J. I. Molinder, *Fundamentals of Signals and Systems—A Building Block Approach,* Cambridge, UK: Cambridge University Press, 2006.

3. C. T. Chen, *Signals and Systems, 3rd ed,* Oxford, UK: Oxford University Press, 2004.

4. J. A. Stuller, *An Introduction to Signals and Systems,* Toronto, CA: Thomson, 2008.

5. E. Kudeki and D. C. Munson, *Analog Signals and Systems,* Upper Saddle River, NJ: Prentice Hall, 2009.

6. M. J. Roberts, *Signals and Systems—Analysis Using Transform Methods and MATLAB, 2nd ed,* , New York, NY: McGraw Hill, 2012.

Problems

3.1 Introduction

3.3.1. For the following RC circuit, $R = 100\ k\Omega$ and $C = 1\ \mu F$. At time $t = 0\ v_c(0^-) = 0\ V$. Solve for $v_{out}(t)$.

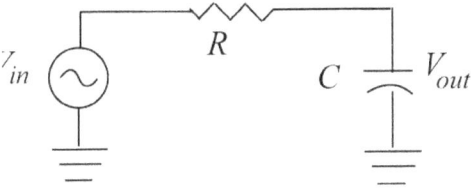

We developed a differential equation for this circuit

$$\frac{dv_o(t)}{dt} + \frac{1}{RC}v_o(t) = \frac{1}{RC}v_{in}(t),$$

or

$$\frac{dv_o(t)}{dt} + 10v_o(t) = 10v_{in}(t).$$

a) Set $v_{in}(t) = \delta(t)$ and solve for the impulse response using Laplace transforms.

b) What is the step response? (The step response is the response to $v_{in}(t) = u(t)$.)

c) Suppose $v_{in}(t) = 0$, but there is an initial charge on the capacitor expressed as $v_o(0^-) = 5V$. What is $v_o(t)$?

3.1.2 Find the Laplace transform of $e^{i\omega t}u(t)$. Use this and the Euler equations to derive the Laplace transforms of $\sin(\omega t)u(t)$ and $\cos(\omega t)u(t)$.

3.1.3. Find the Laplace transform of $\frac{d^2 f(t)}{dt^2}$ given that $\mathcal{L}\{f(t)\} = F(s)$. (Hint: Do not do any integrals. Use the fact that you know the answer for the single derivative.)

3.1.4. Write the Laplace transform of the following equation:
$$\frac{d^2 y(t)}{dt^2} + 5\frac{dy(t)}{dt} + 4y(t) = f(t).$$
If $f(t) = u(t)$, $y(0^-) = 1$, and $y'(0^-) = 2$, solve for Y(s) in the Laplace domain

3.2 Properties of Laplace Transforms

3.2.1. Use the time shifting property and find the Laplace transforms of the following signals:

a. $f(t) = (t-4)^2 u(t-4)$

b. $f(t) = t^2 u(t-4)$

c. $f(t) = t^2 e^{-5t} u(t-3)$

3.2.2. Take the Laplace transforms of the following functions:

a. $f_1(t) = \int_0^t e^{-4\tau} d\tau$

b. $f_2(t) = \int_0^t \tau e^{-2(t-\tau)} d\tau$

c. $f_3(t) = \int_0^t e^{-4\tau} \sin(2(t-\tau)) d\tau$

d. $f_4(t) = te^{-2t} u(t-1)$

e. $f_5(t) = 3t \cos(3t) u(t)$

f. $f_6(t) = t^2 e^{-5t} u(t-1)$

g. $f_7(t) = \int_0^t \tau^2 e^{-2\tau} d\tau$

3.2.3. The inverse Laplace transform of the function

$$F(s) = \frac{1}{(s+1)(s+2)^4}$$

will contain a function of the form $Ae^{-2t} u(t)$. Find A.

3.2.4 Solve the following integral using Laplace transforms:

$$x(t) = \int_0^t (t-\gamma) e^{-5\gamma} d\gamma.$$

3.2.5 Find the Laplace transform of each of the following:

a. $f_1(t) = t^2 e^{-5t} \sin(10^3 t) u(t)$

b. $f_2(t) = (t-2) e^{-(t-4)} u(t)$

3.3 Solving Differential Equations with Laplace Transforms

3.3.1. Take the inverse Laplace transforms of the following functions:

a. $\dfrac{s^2 - 1}{s^2 + 4s + 4}$

b. $\dfrac{e^{-2s}}{s^2 + 5s + 4}$

3.3.2. Solve for B, C and D:

$$X(s) = \frac{s+2}{s(s+1)^4} = \frac{A}{s} + \frac{B}{(s+1)^4} + \frac{C}{(s+1)^3} + \frac{D}{(s+1)^2} + \frac{E}{(s+1)}$$

3.3.3 Solve for D:

$$F_1(s) = \frac{2}{(s+1)(s+2)^3} = \frac{A}{(s+1)} + \frac{B}{(s+2)^3} + \frac{C}{(s+2)^2} + \frac{D}{(s+2)}$$

3.3.4. Solve the following differential equation using Laplace transforms

$$\frac{d^2 y(t)}{dt^2} + 3 \frac{dy(t)}{dt} + 2 y(t) = \frac{df(t)}{dt} + 4 f(t)$$

$f(t) = u(t) \quad y(0^-) = 1, \; y^{(1)}(0^-) = 2$

3.3.5. Find the inverse Laplace transform of the following function:

$$F(s) = \frac{s^2 + 7s + 12}{s^2 + 3s + 2}$$

3.3.6. Find the inverse Laplace transforms of the following functions.

(a) $F_1(s) = \dfrac{s^2 + 3s + 2}{s^2 + 2s + 1}$

(b) $F_2(s) = e^{-s} - e^{-2s}$

(c) $F_3(s) = \dfrac{2s + 3 \times 10^4}{s^2 + 9 \times 10^4 s + 2 \times 10^9}$

3.3.7 Find the Laplace transform of the following function:
$$f(t) = t^3 \cos(10^3 t) u(t)$$

3.3.8 Solve the following for y(t):

$$\dfrac{d^2 y(t)}{dt^2} + 7 \dfrac{dy(t)}{dt} + 12 y(t) = u(t), \quad y(0^-) = 1, \quad y^{(1)}(0^-) = -3.$$

3.3.9 Solve for y(t) using Laplace transforms:

$$2 \dfrac{d^2 y(t)}{dt^2} + 10 \dfrac{dy(t)}{dt} + 8 y(t) = f(t)$$
$$f(t) = e^{-4t} u(t), \quad y(0^-) = -4, \quad y^{(1)}(0^-) = 2$$

3.4 Finding the Inverse of Complex Roots

3.4.1 Solve for y(t):

$$\dfrac{d^2 y(t)}{dt^2} + 2 \dfrac{dy(t)}{dt} + 5 y(t) = -9 f(t)$$
$$y(0^-) = 1, \quad y^{(1)}(0^-) = 0, \quad f(t) = e^{-3t} u(t)$$

3.4.2. Find the step response of the system described by the following equation.

$$\frac{d^2y(t)}{dt^2} + 3\frac{dy(t)}{dt} + 3y(t) = f(t)$$

3.4.3. Solve the following equation for $y(t)$

$$\frac{d^2y(t)}{dt^2} + 4\frac{dy(t)}{dt} + 3y(t) = f(t)$$

$$y(0) = 0, \quad y^{(1)}(0) = 1, \quad f(t) = \sin(2t)u(t)$$

3.4.4. An RCL series circuit has the transfer function

$$H(s) = \frac{R}{sL + 1/sC + R} = \frac{sR/L}{s^2 + sR/L + 1/LC}$$

where $R = 1k\Omega$, $L = 1\,mH$, $C = 1nF$.
Find the impulse response and the step response.

3.4.5. Find the zero-state response.

$$\frac{d^2y(t)}{dt^2} + 4y(t) = \frac{df(t)}{dt} + 3f(t), \quad f(t) = \delta(t).$$

3.4.6. Find the inverse Laplace transform of the following:

$$F_1(s) = \frac{s^2}{s^2 + 10s + 4}$$

3.4.7. Solve for the following system at rest:

$$\frac{d^2y(t)}{dt^2} + 3\times10^4 \frac{dy(t)}{dt} + 4\times10^8 y(t) = 10^8 u(t)$$

3.5 Block Diagrams

3.5.1. Find the total transfer function of the following block diagram:

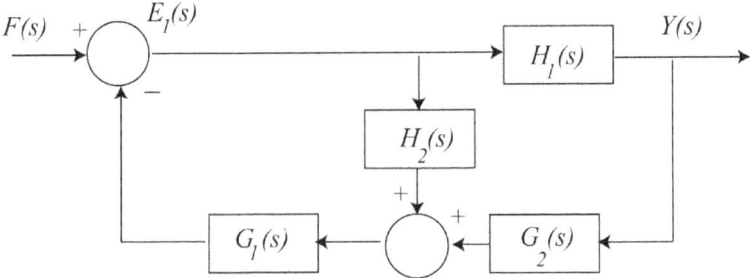

3.5.2. Find the over-all transfer function of the following block diagram:

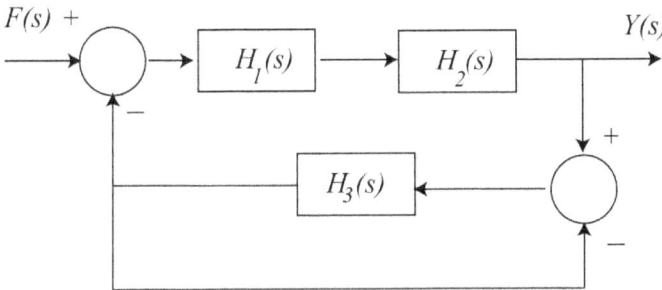

3.5.3. Find the overall transfer function for the following block diagram.

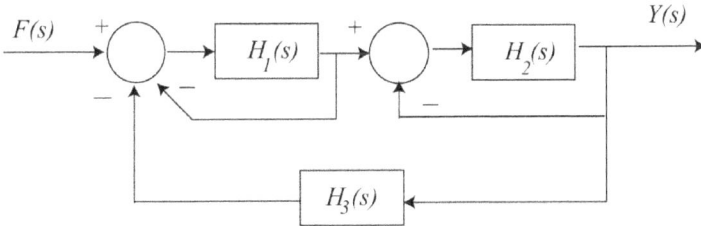

3.5.4 Find the transfer function of the following block diagram.

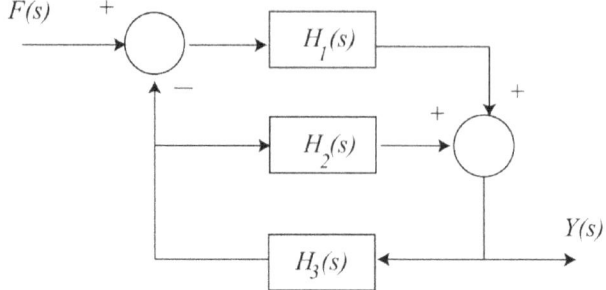

3.5.5 In Fig. 3.15, the system represented by H_1 is described by the following differential equation

$$x''(t)+10^3 x'(t)+10^5 x(t)= f'(t)+10^2 f(t),$$

While the system represented by H_2 is described by

$$y''(t)+3\times 10^2 y'(t)+4\times 10^5 y(t)= x(t).$$

What is the overall transfer function of the following system?

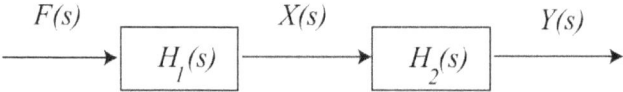

3.5.6. Write the transfer function of the following block diagram.

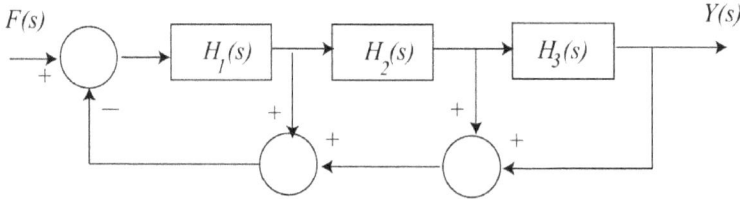

3.5.7. Find the overall transfer function of the following block diagram:

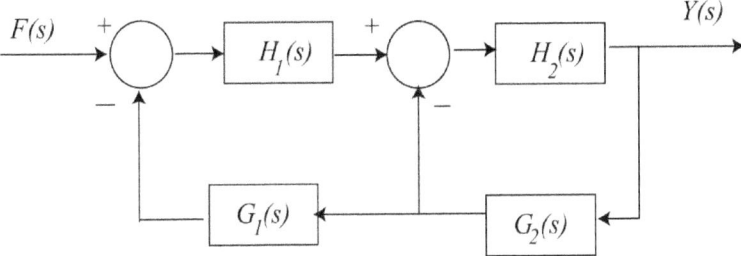

Chapter 4. Fourier Theory

In the chapter on Laplace transforms, we did not concern ourselves too much with either the mathematics or its source. We simply found that the Laplace transform was very useful in solving differential equations and it also gave us a very attractive alternative to convolution integrals.

Now we move to the Fourier transform. It, too, has a forward and inverse transform that allows us to move from the time to the frequency domain and back again. It will not prove particularly useful in solving problems than the methods that we already know. However, the Fourier theory doe have many useful applications in engineering and science, which will become apparent later in this chapter and in following chapters.

4.1. Introduction to Fourier Transforms

The Fourier transform is given by

$$\mathcal{F}\{x(t)\} = X(\omega) = \int_{-\infty}^{\infty} x(t) e^{-j\omega t} dt, \qquad (4.1\ a)$$

and its inverse is

$$\mathcal{F}^{-1}\{X(\omega)\} = x(t) = \frac{1}{2\pi} \int_{-\infty}^{\infty} X(\omega) e^{j\omega t} d\omega. \qquad (4.1\ b)$$

Notice that the forward and inverse Fourier transforms are almost identical. In fact, they differ in only two respects:

1. The forward transform has a term $e^{-j\omega t}$, while the inverse has a term $e^{j\omega t}$.
2. The inverse transform has a term $1/2\pi$ in front of the integral.

4.1.1. Fourier Transforms and the Delta Function

The Fourier transform of the delta function is easy to calculate:

$$\mathcal{F}\{\delta(t)\} = \int_{-\infty}^{\infty} \delta(t) e^{-j\omega t} dt = 1.$$

The inverse Fourier transform has to take the Fourier domain expression back to the time domain:

$$\mathcal{F}^{-1}\{1\} = \frac{1}{2\pi} \int_{-\infty}^{\infty} 1 e^{j\omega t} dt = \delta(t).$$

Similarly, if we take the inverse Fourier transform of the delta function

$$\mathcal{F}^{-1}\{\delta(\omega)\} = \frac{1}{2\pi}\int_{-\infty}^{\infty}\delta(\omega)e^{j\omega t}dt = \frac{1}{2\pi};$$

therefore it must be that

$$\mathcal{F}\left\{\frac{1}{2\pi}\right\} = \int_{-\infty}^{\infty}\frac{1}{2\pi}e^{-j\omega t}dt = \delta(\omega)$$

Actually, in the table of Fourier transforms this pair is written as,

$$1 \Leftrightarrow 2\pi\delta(\omega).$$

Now look at the shifted delta function:

$$\mathcal{F}\{\delta(t-t_0)\} = \int_{-\infty}^{\infty}\delta(t-t_0)e^{-j\omega t}dt = e^{-j\omega t_0}.$$

Therefore,

$$\mathcal{F}^{-1}\{e^{-j\omega t_0}\} = \frac{1}{2\pi}\int_{-\infty}^{\infty}e^{-j\omega t_0}e^{j\omega t}dt = \delta(t-t_0).$$

taking the inverse of a shifted delta in the frequency domain

$$\mathcal{F}^{-1}\{2\pi\delta(\omega-\omega_0)\} = \frac{1}{2\pi}\int_{-\infty}^{\infty}2\pi\delta(\omega-\omega_0)e^{j\omega t}d\omega$$

$$= e^{j\omega_0 t},$$

we can surmise that

$$\mathcal{F}\{e^{j\omega_0 t}\} = 2\pi\delta(\omega-\omega_0).$$

This is a very important result: The Fourier transform of a complex exponential at one frequency gives a delta function in the Fourier domain at that frequency.

4.1.2 Properties of Fourier Transforms

At this point we will stop and look at some of the properties of the Fourier transform.

Property 1. Linearity

$$\mathcal{F}\{\alpha x_1(t) + \beta x_2(t)\} = \alpha X_1(\omega) + \beta X_2(\omega). \qquad (4.2)$$

Proof
The proof comes directly from the definition of the Fourier transform:

$$\mathcal{F}\{\alpha x_1(t)+\beta x_2(t)\} = \int_{-\infty}^{\infty}(\alpha x_1(t)+\beta x_2(t))e^{-j\omega t}dt$$
$$= \alpha \mathcal{F}\{x_1(t)\}+\beta \mathcal{F}\{x_2(t)\}$$

Using this, we can identify other transform pairs relatively easily.

Example 4.1. Find the Fourier transform of the cosine and sine functions.

Solution

$$\mathcal{F}\{\cos(\omega_0 t)\} = \mathcal{F}\left\{\frac{e^{j\omega_0 t}+e^{-j\omega_0 t}}{2}\right\}$$
$$= \frac{1}{2}\left[2\pi\delta(\omega+\omega_0)+2\pi\delta(\omega-\omega_0)\right]$$
$$= \pi\left[\delta(\omega+\omega_0)+\delta(\omega-\omega_0)\right]$$

$$\mathcal{F}\{\sin(\omega_0 t)\} = \mathcal{F}\left\{\frac{e^{j\omega_0 t}-e^{-j\omega_0 t}}{2j}\right\}$$
$$= j\pi\left[\delta(\omega+\omega_0)-\delta(\omega-\omega_0)\right]$$

Property 2. Time Shifting
$$\mathcal{F}\{x(t-t_0)\} = e^{-j\omega t_0} X(\omega). \tag{4.3}$$

Proof

In the Fourier transform definition, Eq. (4.1.1a), change parameters to $\sigma = t - t_0$, and the integral becomes

$$\mathcal{F}\{x(t-t_0)\} = \int_{-\infty}^{\infty} x(\sigma)e^{-j\omega(\sigma+t_0)}d\sigma$$
$$= e^{-j\omega t_0}\int_{-\infty}^{\infty} x(\sigma)e^{-j\omega\sigma}d\sigma = e^{-j\omega t_0} X(\omega)$$

Property 3. Derivative Property
$$\mathcal{F}\left\{\frac{dx(t)}{dt}\right\} = j\omega X(\omega). \tag{4.4}$$

Proof

This time, we start with the inverse definition,

$$x(t) = \frac{1}{2\pi}\int_{-\infty}^{\infty} X(\omega)e^{j\omega t}\,d\omega.$$

Now take the derivative of both sides with respect to t:

$$\frac{d}{dt}x(t) = \frac{d}{dt}\frac{1}{2\pi}\int_{-\infty}^{\infty} X(\omega)e^{j\omega t}\,d\omega$$

$$= \frac{1}{2\pi}\int_{-\infty}^{\infty} j\omega X(\omega)e^{j\omega t}\,d\omega.$$

If this is true, then it must be that

$$\mathcal{F}\left\{\frac{dx(t)}{dt}\right\} = j\omega X(\omega).$$

Repeated use of this technique gives,

$$\mathcal{F}\left\{\frac{d^n x(t)}{dt^n}\right\} = (j\omega)^n X(\omega).$$

Example 4.2: Find the Fourier transform of the sign function

$$\operatorname{sgn}(t) = \begin{cases} 1 & t > 0 \\ 0 & t = 0 \\ -1 & t < 0 \end{cases}$$

Solution

Start by taking the derivative of the *sgn* function:

$$\frac{d}{dt}\operatorname{sgn}(t) = 2\delta(t).$$

The Fourier transform of this function is

$$\mathcal{F}\left\{\frac{d}{dt}\operatorname{sgn}(t)\right\} = 2.$$

But we know that

$$\mathcal{F}\left\{\frac{dx(t)}{dt}\right\} = j\omega X(\omega) = 2,$$

so it must be that

$$\mathcal{F}\{\operatorname{sgn}(t)\} = \frac{2}{j\omega}.$$

Example 4.3. Find the Fourier transform of the Heaviside function

$$u_h(t) = \frac{1}{2} + \frac{1}{2}\operatorname{sgn}(t).$$

Solution

Using the transforms of the constant and of the *sgn* function,

$$\mathcal{F}\{u_h(t)\} = \mathcal{F}\left\{\frac{1}{2} + \frac{1}{2}\operatorname{sgn}(t)\right\} = \pi\delta(\omega) + \frac{1}{j\omega}.$$

Example 4.4. Find the Fourier transform of the following causal function:

$$x(t) = e^{-\alpha t} u(t).$$

Solution

$$\mathcal{F}\{e^{-\alpha t} u(t)\} = \int_0^\infty e^{-\alpha t} e^{-j\omega t}\, dt = \frac{-1}{(\alpha + j\omega)} e^{-\alpha t} e^{-j\omega t} \Big|_0^\infty$$

$$= \frac{-1}{(\alpha + j\omega)}(0 - 1) = \frac{1}{\alpha + j\omega}.$$

Example 4.5

Solve the following using Fourier transforms,

$$\frac{dy(t)}{dt} + 0.5y(t) = \sin(3t)$$

Solution

Take the Fourier transform of both sides

$$j\omega Y(\omega) + 0.5 Y(\omega) = j\pi[\delta(\omega + 3) - \delta(\omega - 3)]$$

Solving for $Y(j\omega)$,

$$Y(\omega) = \frac{j\pi[\delta(\omega+3)-\delta(\omega-3)]}{j\omega+0.5}$$

$$= \frac{j\pi\delta(\omega+3)}{j\omega+0.5} - \frac{j\pi\delta(\omega-3)}{j\omega+0.5}$$

$$= \frac{j\pi\delta(\omega+3)}{-j3+0.5} - \frac{j\pi\delta(\omega-3)}{j3+0.5}$$

This last step occurs because the delta function $\delta(\omega+3)$ has a value only at $\omega=-3$ and $\delta(\omega+3)$ has a value only at $\omega=-3$. Now we write

$$Y(\omega) = \frac{j\pi\delta(\omega+3)}{3.04\angle-80°} - \frac{j\pi\delta(\omega-3)}{3.04\angle 80°}$$

$$= 0.33e^{j80°} j\pi\delta(\omega+3) - 0.33e^{-j80°} j\pi\delta(\omega-3).$$

Taking the inverse Fourier transform we get

$$y(t) = 0.33e^{j80°} j\frac{e^{-j3t}}{2} - 0.33e^{-j80°} j\frac{e^{j3t}}{2}$$

$$= .33\left[\frac{-e^{-j(3t-80°)} + e^{j(3t-80°)}}{2j}\right]$$

$$= 0.33\sin(3t-80°).$$

What would we have found if we had used the old method of phasors from Chapter 1? The phasor equation is,

$$j\omega Y(\omega) + 0.5Y(\omega) = 1,$$

$$\left.\frac{1}{j\omega+0.5}\right|_{\omega=3} = \frac{1}{3.04\angle 80°} = 0.33\angle-80°,$$

$$y(t) = 0.33\sin(3t-80°).$$

Cleary, just using phasors is a much simpler approach.

Property 4. Time Convolution

The Fourier transform of the convolution of two time domain functions is the multiplication of the Fourier transforms of the individual functions:

$$\mathcal{F}\{x_1(t) * x_2(t)\} = X_1(\omega)X_2(\omega). \qquad (4.5)$$

Proof

$$\mathcal{F}\{x_1(t) * x_2(t)\} = \int_{-\infty}^{\infty} \left[\int_{-\infty}^{\infty} x_1(t-\tau)x_2(\tau)d\tau\right] e^{-j\omega t} dt$$

$$= \int_{-\infty}^{\infty} \left[\int_{-\infty}^{\infty} x_1(t-\tau)e^{-j\omega t} dt\right] x_2(\tau) d\tau$$

$$= \int_{-\infty}^{\infty} X_1(\omega)e^{-j\omega\tau} x_2(\tau) d\tau$$

$$= X_1(\omega)X_2(\omega)$$

i.e., convolution in the time domain gives multiplication in the frequency domain.

Example 4.6: Solve using the impulse response:
$$\frac{dy(t)}{dt} + 0.5y(t) = \sin(3t).$$

Solution

To find the impulse response, we replace the forcing function with the delta function:
$$\frac{dy(t)}{dt} + 0.5y(t) = \delta(t).$$
Then take the Fourier transform
$$j\omega Y(\omega) + 0.5Y(\omega) = 1,$$
that gives
$$Y(\omega) = \frac{1}{0.5 + j\omega}.$$
Since this was the response to an impulse input, it is the transfer function
$$H(\omega) = \frac{1}{0.5 + j\omega}.$$
The Fourier transform of the input is
$$X(\omega) = \pi[\delta(\omega+3) - \delta(\omega-3)],$$
so the output is

$$Y(\omega) = H(\omega)X(\omega) = \frac{\pi[\delta(\omega+3) - \delta(\omega-3)]}{0.5 + j\omega}.$$

Clearly this will lead to the same answer as Example 4.5.

Example 4.7. Solve
$$\frac{d^3y(t)}{dt^3} + 2\frac{d^2y(t)}{dt^2} + \frac{dy(t)}{dt} + 3y(t) = \sin(t + 10°) + \cos(2t + 45°)$$

Solution

The transfer function is
$$H(\omega) = \frac{1}{(j\omega)^3 + 2(j\omega)^2 + j\omega - 3} = \frac{1}{3 - 2\omega^2 + j(-\omega^3 + \omega)}$$

We will have to evaluate this at two different frequencies. At $\omega = 1$
$$H(1) = \frac{1}{3 - 2 + j(-1+1)} = 1$$

At $\omega = 2$,
$$H(2) = \frac{1}{3 - 2\cdot 4 + j(-8+2)}$$
$$= \frac{1}{-5 - j6} = \frac{-1}{7.8\angle 50°} = -0.128\angle -50°;$$

so, the result is
$$y(t) = \sin(t + 10°) - 0.128\cos(2t + 45 - 50°)$$
$$= \sin(t + 10°) - 0.128\cos(2t - 5°).$$

Example 4.8. Find the Fourier transform of $p_\tau^h(t)$.

Solution

$$\mathcal{F}\{p_\tau^h(t)\} = \int_{-\tau/2}^{\tau/2} e^{-j\omega t} dt = \frac{1}{-j\omega}\left[e^{-j\omega\tau/2} - e^{j\omega\tau/2}\right]$$

$$= \frac{2}{\omega}\sin\left(\frac{\omega\tau}{2}\right)$$

We would like to write this as a sinc function, which is defined as

$$\text{sinc}(\alpha t) = \frac{\sin(\alpha t)}{\alpha t}. \qquad (4.6)$$

So, we can write

$$\frac{2}{\omega}\sin\left(\frac{\omega\tau}{2}\right) = \tau \cdot \frac{\sin\left(\frac{\omega\tau}{2}\right)}{\left(\frac{\omega\tau}{2}\right)} = \tau \cdot \text{sinc}\left(\frac{\omega\tau}{2}\right).$$

Property 5. Integration

$$\mathcal{F}\left\{\int_{-\infty}^{t} x(\tau)d\tau\right\} = \pi X(0)\delta(\omega) + \frac{1}{j\omega}X(\omega). \qquad (4.7)$$

Proof

Integration can be viewed as the convolution with the Heaviside function:

$$\int_{-\infty}^{t} x(\tau)d\tau = x(t) * u_h(t)$$

Therefore

$$\mathcal{F}\left\{\int_{-\infty}^{t} x(\tau)d\tau\right\} = F\{x(t)*u(t)\} = X(\omega)\left(\pi\delta(\omega) + \frac{1}{j\omega}\right)$$

$$= \pi X(\omega)\delta(\omega) + \frac{1}{j\omega}X(\omega) = \pi X(0)\delta(\omega) + \frac{1}{j\omega}X(\omega).$$

Example 4.9. Find the Fourier Transform of $p_\tau^h(t)$ (assuming we do not already know it).

Solution

We know that

$$\frac{d}{dt}p_\tau^h(t) = \delta\left(t+\frac{\tau}{2}\right) - \delta\left(t-\frac{\tau}{2}\right)$$

so

$$\mathcal{F}\left\{\frac{d}{dt}p_\tau^h(t)\right\} = e^{j\omega\tau/2} - e^{-j\omega\tau/2} = 2j\sin(\omega\tau/2).$$

Using the integration property, we get:

$$\mathcal{F}\{p_\tau^h(t)\} = \frac{1}{j\omega}2j\sin(\omega\tau/2) + \pi 2j\sin(\omega\tau/2)\delta(\omega)$$

$$= \frac{\sin(\omega\tau/2)}{\omega/2}$$

Note: L'Hopital's Rule states that if $g(x)$ is indeterminate at a, i.e., $g(a)=0$, then

$$\lim_{x \to a} \frac{f(x)}{g(x)} = \frac{f'(a)}{g'(a)}.$$

This rule often comes in handy in determining, $X(0)\delta(\omega)$, although we did not need it for this example.

Property 6. Time Multiplication

$$\mathcal{F}\{t^n \cdot x(t)\} = (j)^n \frac{d^n}{d\omega^n} X(\omega) \qquad (4.8)$$

Proof

Start with the definition of the Fourier transforms:

$$X(\omega) = \int_{-\infty}^{\infty} x(t) e^{-j\omega t} dt.$$

Take the derivative of both sides with respect to ω:

$$\frac{d}{d\omega} X(\omega) = \int_{-\infty}^{\infty} (-jt) x(t) e^{-j\omega t} dt,$$

$$j\frac{d}{d\omega} X(\omega) = \int_{-\infty}^{\infty} tx(t) e^{-j\omega t} dt.$$

Higher orders proceed in the same manner.

Example 4.10. Find $\mathcal{F}\{te^{-at}u(t)\}$.

Solution

$$\mathcal{F}\{e^{-at}u(t)\} = \frac{1}{j\omega + a}.$$

$$\mathcal{F}\{te^{-at}u(t)\} = (j)\frac{d}{d\omega}\frac{1}{j\omega + a} = (j)\frac{-j}{(j\omega + a)^2} = \frac{1}{(j\omega + a)^2}.$$

Note: Because this is a causal function we could have looked at the table of Laplace transforms to find

$$\mathcal{L}\{te^{-at}u(t)\} = \frac{1}{(s+a)^2},$$

and we can just replace s with $j\omega$.

Property 7. Frequency Convolution

$$\mathcal{F}\{x_1(t)x_2(t)\} = \frac{1}{2\pi}X_1(\omega) * X_2(\omega) \qquad (4.9)$$

Proof

$$\mathcal{F}^{-1}\{X_1(\omega) * X_2(\omega)\} = \frac{1}{2\pi}\int_{-\infty}^{\infty}\left[\int_{-\infty}^{\infty}X_1(\omega - \lambda) \cdot X_2(\lambda)d\lambda\right]e^{j\omega t}d\omega$$

$$= \frac{1}{2\pi}\int_{-\infty}^{\infty}X_2(\lambda)\left[\int_{-\infty}^{\infty}X_1(\omega - \lambda)e^{j\omega t}d\omega\right]d\lambda$$

$$= \frac{1}{2\pi}\int_{-\infty}^{\infty}X_2(\lambda)x_1(t)e^{j\lambda t}d\lambda = x_1(t)x_2(t)$$

$$\mathcal{F}\{x_1(t) * x_2(t)\} = X_1(\omega) \cdot X_2(\omega)$$

Property 8. Time Reversal

$$X(-\omega) = \int_{-\infty}^{\infty}x(-t)e^{-j\omega t}dt \qquad (4.10)$$

Proof
  First, change variables
$$\sigma = -t,$$
so the Fourier transform integral becomes

$$\int_{\infty}^{-\infty} x(\sigma)e^{\sigma j\omega} \cdot -d\sigma = \int_{-\infty}^{\infty} x(\sigma)e^{\sigma j\omega}d\sigma$$
$$= \int_{-\infty}^{\infty} x(\sigma)e^{-\sigma j(-\omega)}d\sigma = X(-\omega)$$

Example 4.11. Find $\mathcal{F}\{e^{-|t|}\}$

Solution

The function has two parts.
$$e^{-t} \quad t \geq 0$$
$$e^{t} \quad t \leq 0$$

The Fourier transform of the first part is $\dfrac{1}{1+j\omega}$; using time-reversal,

the Fourier transform of the second part is $\dfrac{1}{1-j\omega}$. So

$$X(j\omega) = \frac{1}{1+j\omega} + \frac{1}{1-j\omega} = \frac{2}{1+\omega^2}.$$

Property 9. Frequency Modulation

a) $\mathcal{F}\{x(t)e^{j\omega_0 t}\} = X(\omega - \omega_0),$ \hfill (4.11 a)

b) $\mathcal{F}\{x(t)\cos(\omega_0 t)\} = \dfrac{1}{2}[X(\omega + \omega_0) + X(\omega - \omega_0)],$ \hfill (4.11 b)

c) $\mathcal{F}\{x(t)\sin(\omega_0 t)\} = \dfrac{j}{2}[X(\omega + \omega_0) - X(\omega - \omega_0)].$ \hfill (4.11 c)

Proof

$$\mathcal{F}\{x(t)e^{j\omega_0 t}\} = \int_{-\infty}^{\infty} x(t)e^{j\omega_0 t} e^{-j\omega t} dt$$
$$= \int_{-\infty}^{\infty} x(t)e^{-j(\omega - \omega_0)t} dt = X(\omega - \omega_0).$$

The following two transforms follow directly using Euler's equations:

$$\mathcal{F}\{x(t)\cos(\omega_0 t)\} = \frac{1}{2}[X(\omega+\omega_0) + X(\omega-\omega_0)]$$

$$\mathcal{F}\{x(t)\sin(\omega_0 t)\} = \frac{j}{2}[X(\omega+\omega_0) - X(\omega-\omega_0)]$$

Example 4.12. Find
$$\mathcal{F}[x(t)\cos(\omega_0 t)].$$

Solution

$$\mathcal{F}[x(t)\cos(\omega_0 t)] = \frac{1}{2\pi}X(j\omega) * \pi[\delta(\omega+\omega_0) + \delta(\omega-\omega_0)]$$

$$= \frac{1}{2}[X(\omega+\omega_0) + X(\omega-\omega_0)]$$

This is illustrated below in Fig. 4.1. Convolving $[\delta(\omega+\omega_0) + \delta(\omega-\omega_0)]$ with the transform of $x(t)$, $X(\omega)$, means $X(\omega)$ will be replicated at $-\omega_0$ and ω_0, with the factor one-half. Note that this is the modulation theorem, a property is used extensively in communication theory.

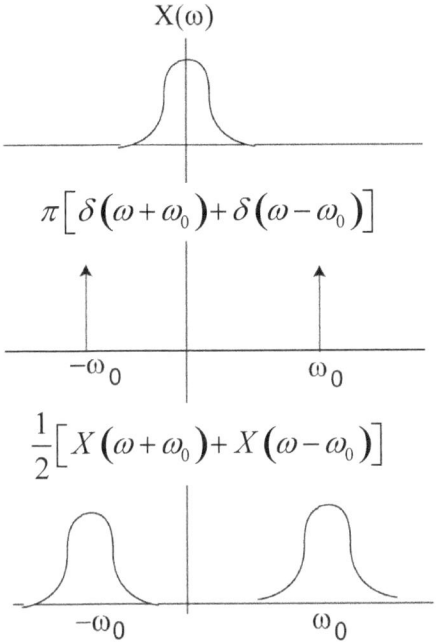

Figure 4.1. $\mathcal{F}[x(t)\cos(\omega_0 t)]$ is the convolution of $X(\omega)$ with Fourier transform of the cosine function, which is two delta function.

Property 10. Duality

If $x(t)$ and $X(\omega)$ are a Fourier transform pair, then the following are also Fourier transform pairs:

$$X(t) \leftrightarrow 2\pi x(-\omega), \qquad (4.12\text{ a})$$
$$X(-t) \leftrightarrow 2\pi x(\omega). \qquad (4.12\text{ b})$$

The duality principle follows from the fact that the forward and inverse Fourier transforms are so similar.

Proof

We start with the Fourier transform integral

and rewrite it as

$$X(\omega) = \int_{-\infty}^{\infty} x(t) e^{-j\omega t} dt,$$

$$X(\lambda) = \int_{-\infty}^{\infty} x(t) e^{-j\lambda t} dt.$$

On the right side of the equation, the variable of integration is t. Now we change to

$$t = -\omega,$$

which will give us

$$X(\lambda) = \int_{\infty}^{-\infty} x(-\omega) e^{j\lambda \omega} \cdot -d\omega$$

$$= \int_{-\infty}^{\infty} x(-\omega) e^{j\lambda \omega} \cdot d\omega.$$

This can be written as

$$X(\lambda) = \frac{1}{2\pi} \int_{-\infty}^{\infty} 2\pi x(-\omega) e^{j\lambda \omega} \cdot d\omega.$$

Now we change variables again

$$\lambda = t,$$

which gives

$$X(t) = \frac{1}{2\pi} \int_{-\infty}^{\infty} 2\pi \cdot x(-\omega) e^{j\omega t} \cdot d\omega.$$

We see that the time-domain signal $X(t)$ and the frequency domain signal $2\pi \cdot x(-\omega)$ are Fourier transform pairs. Similar reasoning would show that

$$X(-t) \leftrightarrow 2\pi x(\omega).$$

This means that for every entry in the Fourier transform table, $x(t) \leftrightarrow X(\omega)$, we can add two more:

$$X(t) \leftrightarrow 2\pi x(-\omega) \text{ and}$$
$$X(-t) \leftrightarrow 2\pi x(\omega).$$

Example 4.13. The first entry in Table 4.1 is,

$$x(t) = \delta(t), \quad X(\omega) = 1,$$

Using duality, find the transform pair that can be added.

Solution

By duality, we can add the entry,
$$X(t)=1 \quad x(-\omega)=2\pi\delta(-\omega)=2\pi\delta(\omega),$$
which is the second entry in Table 4.1

Example 4.14. Find the dual entry for the following transform pair:

$$x(t)=p_\tau(t) \leftrightarrow X(\omega)=\tau\frac{\sin(\omega\tau/2)}{\omega\tau/2}$$

For this example, recall the definition of the sinc function
$$\text{sinc}(\alpha t)=\frac{\sin(\alpha t)}{\alpha t}.$$

Solution

By duality, we also have a term
$$X(t)=\tau\frac{\sin(t\tau/2)}{t\tau/2} \quad 2\pi x(-\omega)=2\pi p_\tau^h(\omega).$$
This is illustrated in Fig. 4.2.

Original: $\quad x(t)=p_\tau^h(t) \quad X(\omega)=\tau\dfrac{\sin(\omega\tau/2)}{\omega\tau/2}$

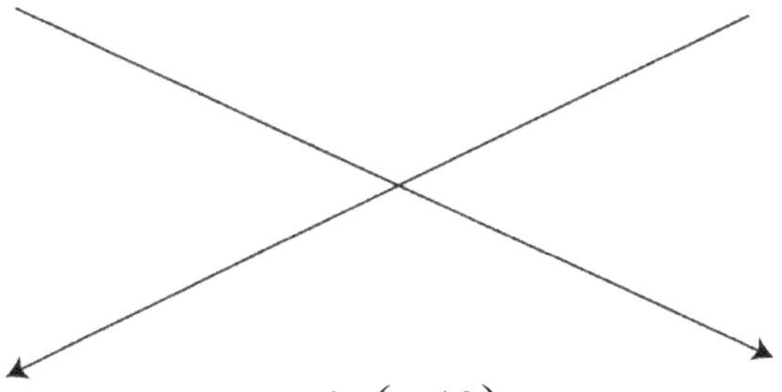

Duality: $\quad X(t)=\tau\dfrac{\sin(t\tau/2)}{t\tau/2} \quad 2\pi x(-\omega)=2\pi p_\tau^h(\omega)$

Figure 4.2. Diagram of the duality principle.

Note that since many of the entries in Table 4.1 are symmetric function, i.e, $x(-t)=x(t)$ and $X(-\omega)=X(\omega)$, the minus signs in the duality definitions, entries (10.a) and (10.b) in Table 4.2, are often not needed.

Example 4.15. Find the Fourier transform of
$$x(t) = \frac{\sin(3t)}{t}.$$

Solution

By duality, we have a transform pair,
$$X(t) = \tau \frac{\sin(t\tau/2)}{t\tau/2} \qquad 2\pi x(-\omega) = 2\pi p_\tau^h(\omega). \qquad (4.13)$$

Now we have to take the given function and write it in the form of $X(t)$ to make the transform. First, we determine τ.
$$x(t) = \frac{\sin(3t)}{t} = \frac{\sin(6t/2)}{t}$$

So obviously we choose $\tau = 6$. Now we can write it in the form of $X(t)$ in Eq. (4.13),
$$x(t) = 3\frac{\sin(6t/2)}{(6t/2)} = \frac{1}{2}\left[6\frac{\sin(6t/2)}{(6t/2)}\right].$$

Using the duality Fourier Transform pair of Eq. (4.1.3),
$$\mathcal{F}\left\{\frac{1}{2}\left[6\frac{\sin(6t/2)}{(6t/2)}\right]\right\} = \frac{1}{2}[2\pi p_\tau(t)] = \pi p_\tau(t).$$

Therefore,
$$\mathcal{F}\left[\frac{\sin(3t)}{t}\right] = \pi p_6^h(\omega).$$

Example 4.16. Find the Fourier transform of
$$g(t) = \frac{1}{3+t^2}.$$

Solution

This is not in the table, but notice the following transforms is
$$x(t) = e^{-\alpha|t|} \Leftrightarrow X(j\omega) = \frac{2\alpha}{\alpha^2 + \omega^2}.$$

Therefore, we can use duality, which tells us we have another entry
$$X(t) = \frac{2\alpha}{\alpha^2 + t^2} \Leftrightarrow x(-\omega) = 2\pi e^{-\alpha|\omega|}. \qquad (4.14)$$

First we have to get $g(t)$ in the form of $\dfrac{2\alpha}{\alpha^2+\omega^2}$:

$$g(t) = \frac{1}{3+t^2} = \frac{1}{2\sqrt{3}}\frac{2\sqrt{3}}{\left(\sqrt{3}\right)^2+t^2},$$

i. e., $\alpha = \sqrt{3}$. Using the duality equation, Eq. (4.1.14),

$$\mathcal{F}\left[\frac{1}{3+t^2}\right] = \frac{1}{2\sqrt{3}}\mathcal{F}\left[\frac{2\sqrt{3}}{\left(\sqrt{3}\right)^2+t^2}\right] = \frac{1}{2\sqrt{3}}\left[2\pi e^{-\sqrt{3}t}\right] = \frac{\pi}{\sqrt{3}}e^{-\sqrt{3}t}$$

Property 11. Parseval's Theorem

Parseval's theorem effectively states that energy is conserved when going back and forth between the time and frequency domains, or

$$E_\infty = \int_{-\infty}^{\infty}|x(t)|^2\,dt = \frac{1}{2\pi}\int_{-\infty}^{\infty}|X(\omega)|^2\,d\omega. \qquad (4.15)$$

Proof
We start by writing the Fourier transform of $x(t)x(t)$ and recall that it can be written as a convolution in the frequency domain (property 7):

$$\int_{-\infty}^{\infty} x(t)x(t)e^{-j\omega t}\,dt = \frac{1}{2\pi}\left[\int_{-\infty}^{\infty} X(\omega-\lambda)\cdot X(\lambda)\,d\lambda\right].$$

At $\omega = 0$,

$$\int_{-\infty}^{\infty} x(t)x(t)\,dt = \frac{1}{2\pi}\left[\int_{-\infty}^{\infty} X(-\lambda)\cdot X(\lambda)\,d\lambda\right].$$

Since $x(t)$ is a real function, we can write

$$\int_{-\infty}^{\infty}|x(t)|^2\,dt = \frac{1}{2\pi}\left[\int_{-\infty}^{\infty}|X(\lambda)|^2\,d\lambda\right].$$

NOTE: then $|X(\lambda)|^2 = X(\lambda)X^*(\lambda)$

Example 4.17. Find the total energy of the signal given by

$$F(\omega) = \frac{6}{9+\omega^2}$$

Solution

First notice that
$$F(\omega) = \frac{2 \cdot 3}{3^2 + \omega^2}$$

The total energy of the above frequency-domain function is
$$E = \frac{1}{2\pi}\int_{-\infty}^{\infty} |F(\omega)|^2 d\omega = \int_{-\infty}^{\infty} |f(t)|^2 dt$$
$$= \int_{-\infty}^{\infty} |e^{-3|t|}|^2 dt = 2\int_{0}^{\infty} e^{-6t} dt$$
$$= \frac{2}{-6} e^{-6t} \Big|_0^{\infty} = \frac{1}{3}.$$

Example 4.18. Evaluate
$$E = \int_{-\infty}^{\infty} \operatorname{sinc}^4(t)\,dt.$$

Solution

We might use Parseval's theorem if we could find the Fourier transform of $\operatorname{sinc}^2(t)$. That is not in our table, but we do have the pair
$$x(t) = \Delta_\tau(t) \quad X(\omega) = \frac{\tau}{2}\operatorname{sinc}^2\left(\frac{\omega\tau}{4}\right).$$

By duality, we have another entry
$$X(-t) = \frac{\tau}{2}\operatorname{sinc}^2\left(\frac{\pi t}{4}\right) \quad 2\pi x(\omega) = 2\pi\Delta_\tau(\omega).$$

If we choose $\tau = 4$
$$X(-t) = 2\operatorname{sinc}^2(t) \quad 2\pi x(\omega) = 2\pi\Delta_4(\omega),$$
or
$$\operatorname{sinc}^2(t) \Leftrightarrow \pi\Delta_4(\omega).$$

Therefore, our integral becomes

$$E = \int_{-\infty}^{\infty} |\sin c(t)|^2 \, dt = \frac{1}{2\pi} \int_{-\infty}^{\infty} |\pi\Delta_4(\omega)|^2 \, d\omega$$

$$= \frac{\pi}{2} \int_{-\infty}^{\infty} |\Delta_4(\omega)|^2 \, d\omega = \pi \int_{0}^{2} \left(1 - \frac{\omega}{2}\right)^2 d\omega$$

$$= \pi \int_{0}^{2} \left(1 - \omega + \frac{\omega^2}{4}\right) d\omega = \pi \left[\omega - \frac{\omega^2}{2} + \frac{1}{3}\frac{\omega^3}{8}\right]_{0}^{2}$$

$$= \pi \left[2 - 2 + \frac{8}{8}\right] = \pi.$$

4.1.3 The Gaussian Function

The Gaussian function plays a key role in mathematics, science and engineering:

$$f(t) = \frac{1}{\sqrt{2\pi}\sigma} e^{-\frac{1}{2}\left(\frac{t}{\sigma}\right)^2}. \tag{4.16 a}$$

The Fourier transform of this function is difficult to calculate, so it will just be given:

$$F(\omega) = e^{-\sigma^2 \omega^2 / 2}. \tag{4.16 b}$$

This is the only function that Fourier transforms to a function of the same shape. Notice something important: The sigma in Eq. (4.16 a) determines how wide the pulse is. However, sigma in the Fourier domain is in the numerator. Thus, the wider the function is in one domain, the narrower it is in the other domain. This is the fabled *Heisenberg Uncertainty Principle*.

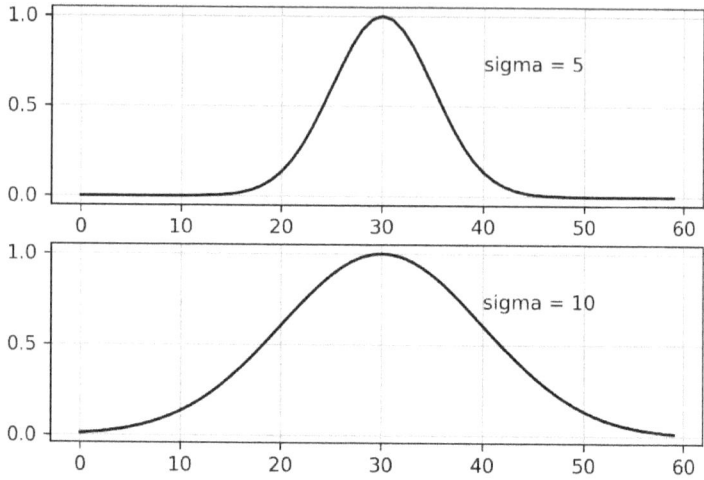

Figure 4.3. The Gaussian function of Eq. (4.16a) for two different values of σ.

Example 4.19. Use duality to verify that both functions of the Gaussian transform pair have the same shape.

Solution
First write the duality entries for the Gaussian pair:

$$F(t) = e^{-\sigma^2 t^2 / 2}$$

$$f(\omega) = 2\pi \frac{1}{\sqrt{2\pi\sigma}} e^{-\frac{1}{2}\left(\frac{\omega}{\sigma}\right)^2} = \frac{\sqrt{2\pi}}{\sigma} e^{-\frac{1}{2}\left(\frac{\omega}{\sigma}\right)^2}$$

Now replace σ with $1/\sigma'$,

$$F(t) = e^{-0.5(t/\sigma')^2} \quad f(\omega) = \sqrt{2\pi}\sigma' e^{-\frac{1}{2}(\sigma'\omega)^2},$$

and divide both sides by $\sqrt{2\pi}\sigma'$ to get

$$F(t) = \frac{1}{\sqrt{2\pi}\sigma'} e^{-t^2/2\sigma'^2} \qquad f(\omega) = e^{-\frac{(\sigma'\omega)^2}{2}}.$$

4.1.4 Using Fourier Transforms to Evaluate Integrals Containing Sinusoids

There are times when the Fourier transform table can be used to evaluate definite integrals extending to infinity.

Example 4.20. Evaluate the following integral

$$I = \int_{-\infty}^{\infty} e^{-\alpha|t|} \cos(\omega_0 t) \, dt$$

Solution

First we use the Euler equations to write

$$I = \int_{-\infty}^{\infty} e^{-\alpha|t|} \left(\frac{e^{j\omega_0 t} + e^{-j\omega_0 t}}{2} \right) dt = \frac{1}{2} \int_{-\infty}^{\infty} e^{-\alpha|t|} e^{j\omega_0 t} \, dt + \frac{1}{2} \int_{-\infty}^{\infty} e^{-\alpha|t|} e^{-j\omega_0 t} \, d.$$

Look at the first integral:

$$I_1 = \frac{1}{2} \int_{-\infty}^{\infty} e^{-\alpha|t|} e^{-j\omega_0 t} \, dt.$$

The equation looks like a Fourier transform, except that it has a definite frequency ω_0. So we can take the Fourier transform, and evaluate at $\omega = \omega_0$

$$I_1 = \frac{1}{2} \mathcal{F}\{e^{-\alpha|t|}\}\Big|_{\omega=\omega_0} = \frac{1}{2} \frac{2\alpha}{\alpha^2 + \omega^2}\Big|_{\omega=\omega_0} = \frac{\alpha}{\alpha^2 + \omega_0^2}.$$

Going back to the original problem, we evaluate for $\omega = -\omega_0$ as well

$$I = 0.5\mathcal{F}\{e^{-\alpha|t|}\}\Big|_{\omega=\omega_0} + 0.5\mathcal{F}\{e^{-\alpha|t|}\}\Big|_{\omega=-\omega_0} = \frac{2\alpha}{\alpha^2 + \omega_0^2}.$$

4.1.5 Review of Section 4.1.

Equations 4.1.1 presented the forward and inverse Fourier transform pair. Recall that the Laplace transform pairs proved very helpful for two reasons:

1. We can use them to solve differential equations.
2. We can use them to do convolution, because a convolution becomes

a multiplication in the Laplace domain.

How about using the Fourier transform? The major difference between the two transforms is that by using Fourier transform, we are dealing with non-causal functions. That means that inputs are sines, cosines, or constants. In every case, that gives us delta functions in the frequency domain. We found out that we could use Fourier transforms to solve differential equations, however it is easier to use the phasor methods introduced in Chapter 1.

We also found that we could do convolutions with Fourier transforms. However, when the input turns into just delta functions, we are only evaluating the transfer function at one frequency:

$$Y(\omega) = H(\omega)\delta(\omega - \omega_1) = H(\omega_1)\delta(\omega - \omega_1).$$

But we already knew how to do that using phasors. Why did we take the time to learn all these properties and do all these transforms back and forth? We will learn that in the next section.

4.2 Fourier Series

4.2.1 Introduction to Inner Products

Recall from analytic geometry, that if we have a vector w in a two-dimensional plane, and we want to write it as the sum of the x and y components,

$$w = a\hat{x} + b\hat{y},$$

we can calculate the coefficients by taking inner products with $\hat{x}$ and $\hat{y}$:

$$w \cdot \hat{x} = a\hat{x} \cdot \hat{x} + b\hat{y} \cdot \hat{x} = a$$
$$w \cdot \hat{y} = a\hat{x} \cdot \hat{y} + b\hat{y} \cdot \hat{y} = b.$$

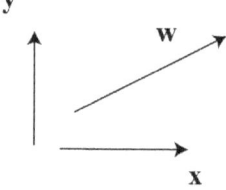

Figure 4.4. A vector in a two-dimensional plane.

We can do this because the unit vectors, $\hat{x}$ and $\hat{y}$, are *orthonormal*. They are orthogonal because

$$\hat{x} \cdot \hat{y} = 0$$

and they are *normalized* because
$$\hat{x} \cdot \hat{x} = \hat{y} \cdot \hat{y} = 1.$$

Over a period of time T, the functions $\sin\left(\frac{2\pi t}{T}\right)$ and $\cos\left(\frac{2\pi t}{T}\right)$ can be considered orthonormal in a similar sense:

$$\frac{2}{T}\int_{-T/2}^{T/2} \sin\left(\frac{2\pi t}{T}\right)\sin\left(\frac{2\pi t}{T}\right) dt = \frac{2}{T}\int_{-T/2}^{T/2} \cos\left(\frac{2\pi t}{T}\right)\cos\left(\frac{2\pi t}{T}\right) dt = 1$$

$$\frac{2}{T}\int_{-T/2}^{T/2} \sin\left(\frac{2\pi t}{T}\right)\cos\left(\frac{2\pi t}{T}\right) dt = 0..$$

In fact, even sines and cosines of different harmonics are orthogonal to each other:

$$\frac{2}{T}\int_{-T/2}^{T/2} \sin\left(\frac{2\pi n t}{T}\right)\sin\left(\frac{2\pi m t}{T}\right) dt = \begin{cases} 1 & \text{if } n = m \\ 0 & \text{if } n \neq m \end{cases},$$

The same property exists for cosines. We will use these concepts to develop the Fourier series.

4.2.2. The Basis of the Fourier Series

Any periodic signal with time period T can be written as a sum of sines and cosines

$$x(t) = \frac{1}{2}a_0 + \sum_{n=1}^{\infty}\left[a_n \cos(n\omega_0 t) + b_n \sin(n\omega_0 t)\right]. \quad (4.17)$$

The fundamental frequency for this time period T is

$$\omega_0 = \frac{2\pi}{T} \frac{\text{radians}}{\text{sec.}} \quad (4.18)$$

The $n=0$ or DC term is

$$a_0 = \frac{2}{T}\int_{-T/2}^{T/2} x(t)\, dt, \quad (4.19\text{ a})$$

and the other terms are

$$a_n = \frac{2}{T}\int_{-T/2}^{T/2} x(t)\cos(n\omega_0 t)\, dt, \quad (4.19\text{ b})$$

$$b_n = \frac{2}{T} \int_{-T/2}^{T/2} x(t)\sin(n\omega_0 t)\, dt. \qquad (4.19\text{ c})$$

The calculation of a_n or b_n is done using the orthogonality properties of sines and cosines, i.e.,

$$\frac{2}{T} \int_{-T/2}^{T/2} \cos(n\omega_0 t)\cos(m\omega_0 t)\, dt = \begin{cases} 1 & \text{if } n = m \\ 0 & \text{if } n \neq m \end{cases}$$

$$\frac{2}{T} \int_{-t/2}^{T/2} \sin(n\omega_0 t)\sin(m\omega_0 t)\, dt = \begin{cases} 1 & \text{if } n = m \\ 0 & \text{if } n \neq m \end{cases}$$

and

$$\frac{2}{T} \int_{-T/2}^{T/2} \sin(n\omega_0 t)\cos(m\omega_0 t)\, dt = 0 \quad \text{for all } n\ \&\ m..$$

The fact that identical functions integrate to one indicates that they are *orthonormal*

If we have a periodic signal $x(t)$ with time period T, then we can write it like Eq. (4.17). We determine the a_n s and b_n s by using the orthonormality properties of the sines and cosines. For instance, we calculate the coefficient a_m by multiplying $x(t)$ by $\cos(m\omega_0 t)$ and then integrating over the time period:

$$a_m = \frac{2}{T}\int_0^T \left[\frac{1}{2}a_0 + \sum_{n=1}^{\infty}\left[a_n\cos(n\omega_0 t) + b_n\sin(n\omega_0 t)\right]\right]\cos(m\omega_0 t)\, dt.$$

The integral of $a_0 \cos(m\omega_0 t)$ over one period T will be zero. Similarly the integral of $\cos(m\omega_0 t)$ with any sine term will be zero, and the integral of $\cos(m\omega_0 t)$ with any other cosine except $m=n$ will be zero. There will only be one term left:

$$a_m = \frac{2}{T}\int_{-T/2}^{T/2} a_m \cos(m\omega_0 t)\cos(m\omega_0 t)\, dt$$

$$= \frac{2}{T} a_m \int_{-T/2}^{T/2} \left[\frac{1}{2} - \frac{1}{2}\cos(2m\omega_0 t)\right] dt$$

$$= \frac{2}{T} a_m \int_{-T/2}^{T/2} \frac{1}{2}\, dt = \frac{2}{T} a_m \frac{1}{2} T = a_m$$

We would obtain a similar result for any of the b terms.

Example 4.21. Calculate the Fourier series for the rectangular series shown in Fig. 4.5.

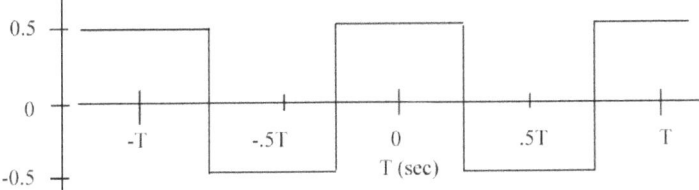

Figure 4.5. A periodic time-domain signal

Solution

We can simplify the calculation in two ways. First of all, we will add a DC term of *1/2*, and then eliminate the calculation of a_0 (Fig. 4.6). Then, using symmetry, we calculate over the interval 0 to T/2, and double it.

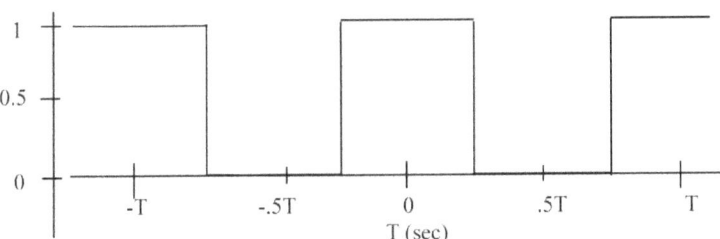

Figure 4.6. The same signal as Fig. 4.5, but shifted upward.

$$a_n = \frac{2}{T} 2 \int_0^{T/4} \cos(n\omega_o t) dt = \frac{4}{T} \left\{ \frac{1}{n\omega_o} \left[\sin(n\omega_o t) \right] \Big|_0^{T/4} \right\}$$

$$= \frac{4}{2\pi n} \left\{ \sin\left(\frac{n2\pi T}{T4}\right) \right\} = \frac{2}{\pi n} \sin\left(\frac{n\pi}{2}\right)$$

$$a_n = \begin{cases} 2/n\pi & n = 1,5,9 \\ -2/n\pi & n = 3,7,11 \\ 0 & n = 0,2,4 \end{cases}$$

Notice that just the first two non-zero terms of the Fourier series result in a good approximation (Fig. 4.7 a). As more terms are added, the series comes closer to the rectangular function (Fig. 4.7 b and 4.7 c).

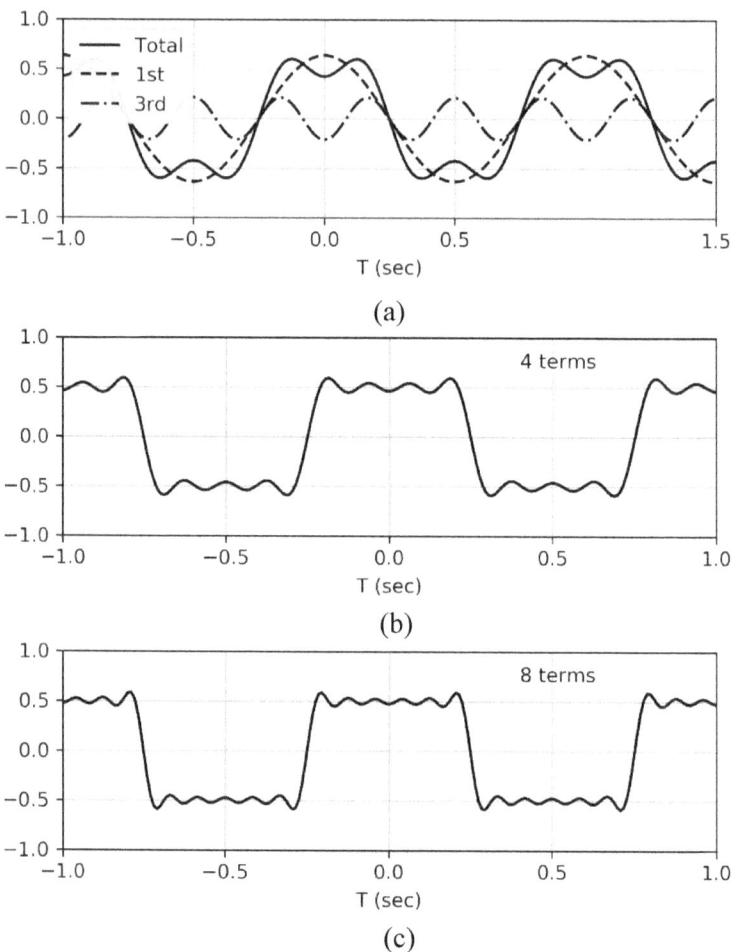

Figure 4.7. (a) Reconstruction of the series of Fig. 4.5 using just the first two nonzero terms; (b) reconstruction from the first four nonzero terms; (c) reconstruction from the first eight nonzero terms.

4.2.3. The Complex Fourier Series

In general, even if we only have one frequency, say $\omega_n = n\omega_0$, we still need two numbers, a_n and b_n, to describe the *nth* series term.

Since we know that the sine and cosine terms are orthonormal, we might wonder if we could change the two real numbers to one complex number. Start with

$$x_n(t) = a_n \cos(n\omega_n t) + b_n \sin(n\omega_n t),$$

and use Euler's equations:

$$x_n(t) = a_n \left[\frac{e^{jn\omega_n t} + e^{-jn\omega_n t}}{2}\right] + b_n \left[\frac{e^{jn\omega_n t} - e^{-jn\omega_n t}}{2j}\right].$$

We begin by grouping the positive and negative frequency components:

$$x_n(t) = \left(\frac{a_n}{2} - j\frac{b_n}{2}\right) e^{jn\omega_n t} + \left(\frac{a_n}{2} + j\frac{b_n}{2}\right) e^{-jn\omega_n t}$$

$$= X_n e^{jn\omega_n t} + X_n^* \left(e^{-jn\omega_n t}\right)$$

The original series

$$x(t) = \frac{1}{2} a_0 + \sum_{n=1}^{\infty} \left[a_n \cos(n\omega_0 t) + b_n \sin(n\omega_0 t)\right]$$

can be written

$$x(t) = \sum_{n=-\infty}^{\infty} X_n e^{jn\omega_0 t}, \qquad (4.20 \text{ a})$$

where

$$X_0 = \frac{a_0}{2}, \quad X_n = \left(\frac{a_n}{2} - j\frac{b_n}{2}\right), \quad X_n^* = \left(\frac{a_n}{2} + j\frac{b_n}{2}\right).$$

Notice that the new series goes between plus and minus infinity because the Euler equations use plus and minus terms. We determine the X_n in a manner similar to the way we calculated a_n and b_n:

$$X_n = \frac{1}{T} \int_{-T/2}^{T/2} x(t) e^{-jn\omega_0 t} dt. \qquad (4.20 \text{ b})$$

This depends on a similar orthonormality condition:

$$\frac{1}{T} \int_{-T/2}^{T/2} e^{jn\omega_0 t} e^{jm\omega_0 t} dt = \frac{1}{T} \int_{-T/2}^{T/2} e^{j(n-m)\omega_0 t} dt = 1 \quad \text{if } n = m,$$

Otherwise,

$$\frac{1}{T}\int_{-T/2}^{T/2} e^{jn\omega_o t} e^{jm\omega_o t}\, dt = \frac{1}{T}\int_{-T/2}^{T/2} e^{j(n-m)\omega_o t}\, dt$$

$$= \frac{1}{T}\frac{1}{j(n-m)\omega_o} e^{j(n-m)\omega_o t}\Big|_{t=-T/2}^{t=T/2}$$

$$= \frac{1}{T}\frac{1}{j(n-m)\omega_o}\left(e^{j(n-m)\pi} - e^{-j(n-m)\pi}\right) = 0.$$

When we try to calculate the *n*th coefficient

$$X_n = \frac{1}{T}\int_0^T \sum_{m=-\infty}^{\infty} X_m e^{jm\omega_o t} e^{-jn\omega_o t}\, dt,$$

the only term that survives is $m = n$,

$$X_n = \frac{1}{T}\int_0^T X_n e^{jn\omega_o t} e^{-jn\omega_o t}\, dt = \frac{1}{T}\int_0^T X_n\, dt = X_n.$$

Note that the complex form has plus and minus values. However, $X_{-n} = X_n^*$; so if we know the positive value, we also know the negative. Once again we can plot the coefficients X_n out, but since they are complex, we need to plot magnitude and phase. This is called the *line spectra*.

Example 4.22. Redo example 4.21 using the complex series.

Solution

As before, it will not hurt to add a DC term, and then leave it out of the series. So now we calculate

$$X_n = \frac{1}{T}\int_{-T/2}^{T/2} x(t) e^{-jn\omega_o t}\, dt$$

$$= \frac{1}{T}\int_{-T/4}^{T/4} e^{-jn\omega_o t}\, dt = \frac{1}{T}\frac{1}{-jn\omega_o} e^{-jn\omega_o t}\Big|_{-T/4}^{T/4}$$

$$= \frac{1}{T}\frac{1}{-jn\omega_o}\left[e^{-jn\omega_o T/4} - e^{jn\omega_o T/4}\right].$$

Remember that

$$\omega_o = \frac{2\pi}{T},$$

so

$$X_n = \frac{1}{T-jn2\pi} \frac{T}{} \left[e^{-jn\pi/2} - e^{jn\pi/2} \right]$$

$$= \frac{1}{-jn2\pi}\left[-2j\sin\left(\frac{n\pi}{2}\right)\right] = \frac{1}{n\pi}\sin\left(\frac{n\pi}{2}\right)$$

$$X_1 = X_{-1} = \frac{1}{\pi}$$

$$X_2 = \frac{1}{n\pi}\sin\left(\frac{2\pi}{2}\right) = 0,$$

as will all even terms. Furthermore,

$$X_3 = X_{-3} = \frac{1}{3\pi}\sin\left(\frac{3\pi}{2}\right) = -\frac{1}{3\pi}$$

Since the positive and negative terms are the same.

$$x(t) = \sum_{n=-\infty}^{\infty} X_n e^{jn\omega_0 t} = \sum_{n=1,3,5}^{\infty} X_n \left(e^{jn\omega_0 t} + e^{-jn\omega_0 t} \right)$$

$$= \sum_{n=1,3,5}^{\infty} 2X_n \cos(n\omega_0 t) = \frac{2}{\pi}\begin{bmatrix} \cos(\omega_0 t) - \frac{1}{3}\cos(3\omega_0 t) \\ + \frac{1}{5}\cos(5\omega_0 t) - ... \end{bmatrix}.$$

Note that even though we used complex functions to calculate the series terms, we grouped them into cosine terms at the end to compose a time domain function that had no imaginary terms.

4.2.4. Fourier Series Using Fourier Transforms

Look at the following periodic signal. The signal $x_0(t)$ repeats very T seconds:

$$x(t) = \sum_{n=-\infty}^{\infty} x_0(t-nT) \qquad (4.21)$$

We saw that this periodic signal can be written as a complex Fourier series

$$x(t) = \sum_{n=-\infty}^{\infty} X_n e^{jn\omega_0 t}, \qquad (4.22\text{ a})$$

where X_n is determined by

$$X_n = \frac{1}{T}\int_{-T/2}^{T/2} x_0(t)e^{-jn\omega_0 t}dt. \quad (4.22\text{ b})$$

Compare this to the Fourier transform:

$$X(\omega) = \int_{-\infty}^{\infty} x_0(t)e^{-j\omega t}dt. \quad (4.22\text{ c})$$

If we only wanted the Fourier transform at one frequency, say $\omega = n\omega_0$, we would have

$$X(n\omega_0) = \int_{-T/2}^{T/2} x_0(t)e^{-jn\omega_0 t}dt.$$

Comparing this with our equation for the calculation of X_n, Eq. (4.22 b):

$$X_n = \frac{1}{T}X(n\omega_0).$$

The point is, if we are looking for the Fourier series of a repeating function, we can just take the Fourier transform of that function, and write down the coefficients corresponding to the harmonic frequencies.

Here is a summary of the procedure:
1. For a signal $x(t)$ that repeats over a time interval, T, take the signal over just one time interval,
$$x_0(t) = x(t), \quad -T/2 \leq t < T/2.$$
2. Use the tables and the related theorems to determine the Fourier transform of the function,
$$X_0(\omega) = \mathcal{F}\{x_0(t)\}.$$
3. Get the complex series coefficients by evaluating at the harmonic frequencies and dividing by the time interval
$$X_n = \frac{1}{T}X_0(n\omega_0),$$
where
$$\omega_0 = \frac{2\pi}{T}.$$
4. If the coefficients are even, $X_n = X_{-n}$, the series can be simplified to a cosine series; if the coefficients are odd, $X_n = -X_{-n}$, the series can be simplified to a sine series. Sometimes both a cosine and sine series are needed.

Example 4.23 Find the Fourier series of Example 4.21 using Fourier transforms.

<u>Solution</u>

Step 1. The function over one time period is $x_0(t) = p_{T/2}(t)$; and the fundamental frequency is $\omega_0 = \dfrac{2\pi}{T} = 2\pi$.

Step 2.
$$X_0(\omega) = \tau \dfrac{\sin\left(\dfrac{\omega\tau}{2}\right)}{\dfrac{\omega\tau}{2}} = \dfrac{T}{2}\dfrac{\sin\left(\dfrac{\omega T}{4}\right)}{\dfrac{\omega T}{4}}$$

Step 3.
$$X_n = \dfrac{1}{T}X_0(\omega = n\omega_0) = \dfrac{1}{2}\dfrac{\sin\left(\dfrac{n\pi}{2}\right)}{\dfrac{n\pi}{2}} = \dfrac{\sin\left(\dfrac{n\pi}{2}\right)}{n\pi}$$

Step 4.
We can see that $X_{-n} = X_n$, so
$$x(t) = \sum_{n=1}^{\infty} X_n \left(e^{jn\omega_0 t} + e^{-jn\omega_0 t}\right) = \sum_{n=1}^{\infty} 2X_n \cos(n\omega_0 t)$$
$$= \dfrac{2}{\pi}\left[\cos(\omega_0 t) - \dfrac{1}{3}\cos(3\omega_0 t) + \dfrac{1}{5}\cos(5\omega_0 t) - \ldots\right]$$

Example 4.24. Find the Fourier series of the function in Fig. 4.8.

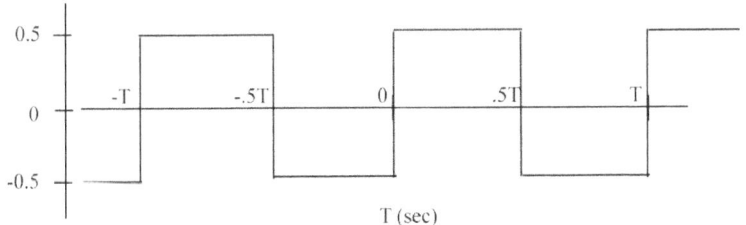

Figure 4.8. A rectangular series with time period T.

Solution

Step 1.
$$x_0(t) = -\frac{1}{2} p_{T/2}(t+T/4) + \frac{1}{2} p_{T/2}(t-T/4),$$
in the interval -0.5 T to 0.5T.

Step 2.
$$X_0(\omega) = \left(-e^{j\omega T/4} + e^{-j\omega T/4}\right)\frac{1}{2}\frac{T}{2}\frac{\sin\left(\frac{\omega T}{4}\right)}{\frac{\omega T}{4}}$$

Step 3.
$$X_n = \frac{1}{T} X(\omega = n\omega_0 = 2\pi n) = \left(-e^{j\pi n/2} + e^{-j\pi n/2}\right)\frac{1}{2}\frac{\sin\left(\frac{n\pi}{2}\right)}{n\pi}$$
$$= -j\sin\left(\frac{\pi n}{2}\right)\frac{\sin\left(\frac{n\pi}{2}\right)}{n\pi} = -\frac{j}{n\pi}\sin^2\left(\frac{n\pi}{2}\right)$$

Step 4.
Since $X_{-n} = -X_n$ we will convert the exponential to the sine series
$$x(t) = \sum_{n=1}^{\infty} X_n\left(e^{jn\omega_0 t} - e^{-jn\omega_0 t}\right) = \sum_{n=1}^{\infty} \frac{2}{n\pi}\sin^2\left(\frac{n\pi}{2}\right)\sin(n\omega_0 t)$$
$$= \frac{2}{\pi}\left[\sin(\omega_0 t) + \frac{1}{3}\sin(3\omega_0 t) + \frac{1}{5}\sin(5\omega_0 t) + ...\right]$$

The bottom line is that we can bring to bear everything we have learned about Fourier transforms to help us calculate the Fourier series.

Example 4.25. Write the Fourier series of the function in Fig. 4.9. (T = 1).

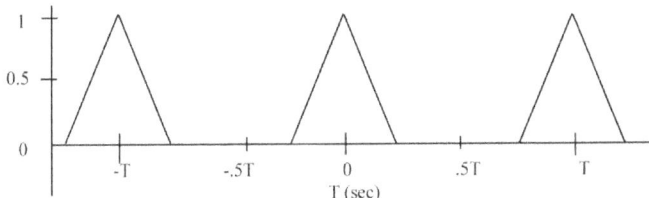

Figure 4.9. A series of triangle functions.

Solution

Step 1 $$x_0(t) = \Delta_{T/2}(t)$$

Step 2 $$\mathcal{F}\{\Delta_{T/2}\} = X(\omega) = \frac{\tau}{2}\text{sinc}^2\left(\frac{\omega\tau}{4}\right) = \frac{1}{4}\text{sinc}^2\left(\frac{\omega}{8}\right)$$

Step 3 $$X_n = \frac{1}{T}X(jn\omega_0) = \frac{1}{4}\text{sinc}^2\left(\frac{n\omega_0}{8}\right)$$

Notice that $X_n = X_{-n}$, and the dc term $X(\omega = 0) = 1/4$ so the series is,

Step 4 $$x(t) = \sum_{n=-\infty}^{\infty} X_n e^{jn\omega_0 t} = \frac{1}{4} + \sum_{n=1}^{\infty} 2X_n \cos(n\omega_0 t)$$

Example 4.26. Find the Fourier series of the function given by

$$x(t) = \sum_{n=-\infty}^{\infty} e^{-2|t-nT|} \quad T = 1$$

Solution

Look only at the term centered at t = 0:
$$x_0(t) = e^{-2|t|}$$
We know that its Fourier transform is
$$X(\omega) = \frac{2 \cdot 2}{2^2 + \omega^2}.$$
So the X_n terms are
$$X_n(\omega) = \frac{1}{T}\frac{4}{4+(n\omega_0)^2} = \frac{4}{4+(2\pi n)^2} = \frac{1}{1+(\pi n)^2}.$$
Obviously, these are even, except for the dc term, which is

151

so we can write,

$$X_0(\omega) = \frac{1}{1+(\pi 0)^2} = 1,$$

$$x(t) = \sum_{n=-\infty}^{\infty} X_n e^{jn\omega_0 t} = 1 + \sum_{n=1}^{\infty} X_n \left(e^{jn\omega_0 t} + e^{-jn\omega_0 t} \right)$$

$$= 1 + \sum_{n=1}^{\infty} 2 X_n \cos(n\omega_0 t).$$

Theorem. The Fourier transform of a real time domain function $x(t)$ has a real part that is symmetric in ω and an imaginary part that is asymmetric in ω.

Proof

$$X(\omega) = \int_{-\infty}^{\infty} x(t) e^{-j\omega t} dt = \int_{-\infty}^{\infty} x(t) \cos(\omega t) dt - j \int_{-\infty}^{\infty} x(t) \sin(\omega t) dt$$

Notice that

$$X_{real}(-\omega) = \int_{-\infty}^{\infty} x(t) \cos(-\omega t) dt = \int_{-\infty}^{\infty} x(t) \cos(\omega t) dt = X_{real}(\omega),$$

while

$$X_{imag}(-\omega) = \int_{-\infty}^{\infty} x(t) \sin(-\omega t) dt = -\int_{-\infty}^{\infty} x(t) \sin(\omega t) dt = X_{imag}(-\omega)$$

Example 4.27. Write the Fourier series of this function:

$$x(t) = \sum_{n=-\infty}^{\infty} e^{-2(t-n)} u(t-n)$$

Solution

Step 1: $\quad x_0(t) = e^{-2t} u(t), \quad T = 1$

Step 2: $\quad X_0(\omega) = \frac{1}{2+j\omega} \frac{2-j\omega}{2-j\omega} = \frac{2-j\omega}{4+\omega^2}, \quad \omega_0 = 2\pi$

Step 3: $X_n(\omega) = \dfrac{2 - j(2\pi n)}{4 + (2\pi n)^2}$, $X_0(\omega) = \dfrac{1}{2}$

Note that the real part is symmetric and the imaginary part is antisymmetric.

$$X_n rl(\omega) = \dfrac{2}{4 + (2\pi n)^2}, \quad X_n im(\omega) = \dfrac{j(2\pi n)}{4 + (2\pi n)^2}$$

Step 4:
$$x(t) = X_0 + \sum_{n=1}^{\infty} X_n rl \cdot 2\cos(2\pi nt) + \sum_{n=1}^{\infty} X_n im \cdot j2\sin(2\pi nt)$$

$$= \dfrac{1}{2} + \sum_{n=1}^{\infty} \dfrac{4}{4 + (2\pi n)^2} \cos(2\pi nt) - \sum_{n=1}^{\infty} \dfrac{4\pi n}{4 + (2\pi n)^2} \sin(2\pi nt)$$

4.3 Fourier Transforms of Fourier Series

In the previous section we saw that any periodic signal can be written

$$x(t) = \sum_{n=-\infty}^{\infty} X_n e^{jn\omega_0 t},$$

and the terms X_n can be determined by

$$X_n = \dfrac{1}{T} \int_{-T/2}^{T/2} x(t) e^{-jn\omega_0 t} dt.$$

Now we take the Fourier Transform of $x(t)$

$$X(\omega) = \sum_{n=-\infty}^{\infty} X_n 2\pi \delta(\omega - n\omega_0).$$

If we run this signal through a linear system described by the transfer function $H(\omega)$, the output in the frequency domain is

$$Y(\omega) = H(\omega) X(\omega) = \sum_{n=-\infty}^{\infty} X_n H(\omega) 2\pi \delta(\omega - n\omega_0)$$

$$= \sum_{n=-\infty}^{\infty} X_n H(n\omega_0) 2\pi \delta(\omega - n\omega_0).$$

(4.23)

Going back to the time domain gives,

$$y(t) = \sum_{n=-\infty}^{\infty} X_n |H(n\omega_0)| e^{j(n\omega_0 t + \angle H(n\omega_0))}. \qquad (4.24)$$

Equation (4.24) shows that the output will be the same frequencies as the input, with amplitudes of each frequency component attenuated and phase shifted, just as if they were a group of individual frequencies.

Suppose instead we had

$$x(t) = \frac{a_0}{2} + \sum_{n=1}^{\infty} \left[a_0 \cos(n\omega_0 t) + b_0 \sin(n\omega_0 t) \right]$$

If we run this through the same linear system, we get

$$y(t) = H(0)\frac{a_0}{2} + \sum_{n=1}^{\infty} |H(n\omega_0)| \left[\begin{array}{l} a_0 \cos(n\omega_0 t + \angle H(n\omega_0)) \\ + b_0 \sin(n\omega_0 t + \angle H(n\omega_0)) \end{array} \right]$$

This tells us the same thing: The frequency components are attenuated and phase shifted according to their frequency.

Example 4.28 Determine the output if the series of square pulses (Fig. 4.10) is put into a system described by the following differential equation

$$\frac{dy(t)}{dt} + 4y(t) = 4x(t).$$

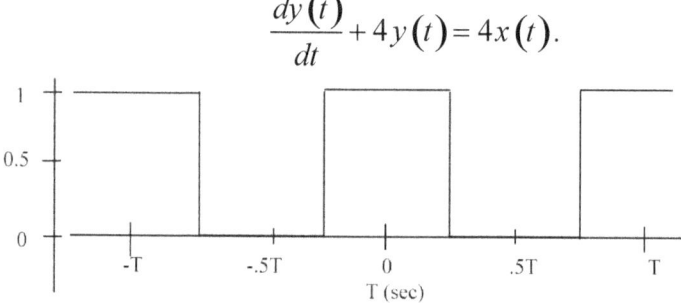

Figure 4.10. The function $x(t)$ is a series of pulses at one second intervals.

Solution

The system has the transfer function

$$H(\omega) = \frac{4}{j\omega + 4}.$$

This is a single pole, low-pass filter. The series in Fig. 4.3.1 differs from Example 4.2.3, only in the DC term. We can see that the average DC value of once cycle is ½, so we write,

$$x(t) = \frac{1}{2} + \frac{2}{\pi}\cos(2\pi t) - \frac{2}{3\pi}\cos(6\pi t) + \dots.$$

To determine the output, it is necessary to determine the value of the transfer function $H(\omega)$ at each of the harmonic frequencies:

$$H(\omega = 2\pi) = \frac{4}{j2\pi + 4} = \frac{4}{7.45\angle \tan^{-1}(2\pi/4)} = 0.54\angle -57°$$

$$H(\omega = 6\pi) = \frac{4}{j6\pi + 4} = \frac{4}{19.3\angle \tan^{-1}(6\pi/4)} = 0.21\angle -78°$$

The output can be approximated by the first three non-zero term

$$y(t) \cong \frac{1}{2} + \frac{2}{\pi}(0.54)\cos(2\pi t - 57°) - \frac{2}{3\pi}(0.303)\cos(6\pi t - 78°).$$

We can see that the higher order harmonics will attenuate rapidly, so these three terms are probably adequate.

Example 4.29. What would be different if the system in the previous example was

$$\frac{d^2y(t)}{dt^2} + 8\frac{dy(t)}{dt} + 16y(t) = 4x(t)?$$

Solution

The transfer function is

$$H(s) = \frac{4}{s^2 + 8s + 16} = \frac{4}{(s+4)^2} = \frac{1}{4}\left(\frac{4}{s+4}\right)^2.$$

Note how this compares with the previous example. Other than the factor ¼, it is two identical low pass filters. Therefore, we can take the results of the previous example, square the attenuations, and double the phase shifts.

Example 4.30. The periodic signal in Fig. 4.11 is put through the circuit in Fig. 4.12. What is the output?

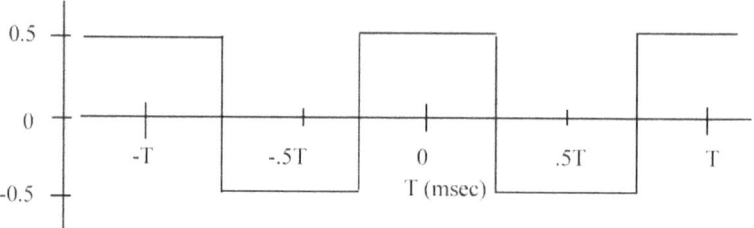

Figure 4.11. A rectangular pulse series with $T = 1$ ms.

Solution

Figure 4.11 contains the same waveform as Example 4.21, except the units are now milliseconds. Therefore, the series is

$$x(t) = \frac{2}{\pi}\left[\begin{array}{l}\cos(\omega_0 t) - \frac{1}{3}\cos(3\omega_0 t) \\ + \frac{1}{5}\cos(5\omega_0 t) - \frac{1}{7}\cos(7\omega_0 t) + ...\end{array}\right].$$

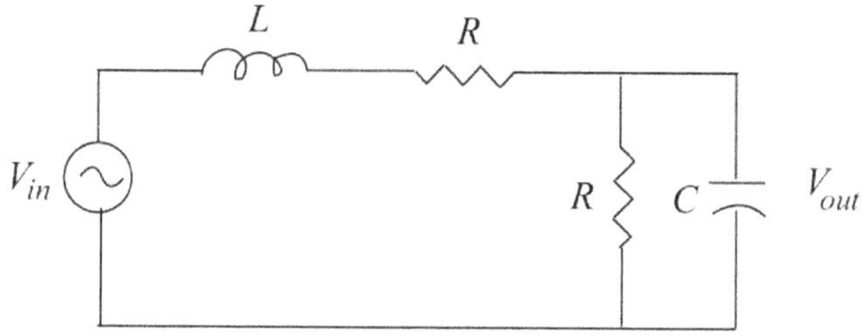

Figure 4.12. An analog circuit with the values $R = 10k\Omega$, $L = 5\,mH$, and $C = 2\,\mu F$

The circuit has the transfer function,

$$H(s) = \frac{1/LC}{s^2 + \left(\dfrac{1}{R_2 C} + \dfrac{R_1}{L}\right)s + \dfrac{R_1 + R_2}{R_2 LC}}$$

$$= \frac{10^8}{(j\omega)^2 + (2.00005 \times 10^6)j\omega + 2 \times 10^8}.$$

At the specific frequencies, $\omega = n\omega_0$:

$$H(n\omega_0) = \frac{10^8}{(2\times 10^8 - n\omega_0^2) + jn\omega_0(2\times 10^6)}.$$

If for instance, $T = 1$ ms,

$$\omega_0 = \frac{2\pi}{10^{-3}} = 2\pi \times 10^3$$

$$H(n\omega_0) = \frac{10^8}{(2\times 10^8 - n(2\pi \times 10^3)^2) + jn(2\pi \times 10^3)(2\times 10^6)}$$

$$= \frac{10^8}{(2\times 10^8 - n4\pi^2 \times 10^6) + jn(4\pi \times 10^9)}$$

The values of H at the first three harmonics are:

$$H(1\omega_0) \cong \frac{10^8}{+j(4\pi \times 10^9)} = 0.008\angle -90^\circ$$

$$H(3\omega_0) \cong \frac{10^8}{+j(12\pi \times 10^9)} = 0.0026\angle -90^\circ$$

$$H(5\omega_0) \cong \frac{10^8}{+j(20\pi \times 10^9)} = 0.0016\angle -90^\circ$$

$$y(t) = \frac{2E}{\pi}\left[\begin{array}{l} 0.008\cos(\omega_0 t - 90^\circ) - \frac{1}{3}(0.0026)\cos(3\omega_0 t - 90^\circ) \\ + \frac{1}{5}(0.016)\cos(5\omega_0 t - 90^\circ) - \ldots \end{array}\right]$$

Example 4.31. An input into a low pass filter is a group of delta functions spaced at one millisecond intervals. The filter is a four-pole low pass filter with a cutoff at 2.5 kHz. What comes out of the filter? Your answer need only be accurate to 10 percent. Assume each delta function has a magnitude of 10^{-3}.

Solution

The delta functions can be expressed mathematically as

$$x(t) = \sum_{n=-\infty}^{\infty} 10^{-3} \delta(t-nT) \quad T = 1\,ms.$$

We start by writing the Fourier series of $x(t)$:

$$x(t) = \sum_{n=-\infty}^{\infty} X_n e^{jn\omega_0 t}$$

Since the function over one interval is $x_0(t) = 10^{-3}\delta(t-nT)$, the coefficients are

$$X_n = \frac{1}{T}\int_{-T/2}^{T/2} 10^{-3}\delta(t)e^{-jn\omega_0 t} = 1,$$

and the fundamental frequency is

$$\omega_0 = \frac{2\pi}{T} = 2\pi 10^3.$$

Therefore, we can write the series as

$$x(t) = 1 + \sum_{n=1}^{\infty} \cos(2\pi 10^3 nt).$$

The cutoff of the filter is at 2.5 kHz. A four-pole filter means that almost anything above the cutoff is eliminated, so we can simply write

$$y(t) = 1 + \sum_{n=1}^{2} \cos(2\pi 10^3 nt).$$

Notice that it is straight forward to calculate what happens to a Fourier series going through a linear system: Simply calculate the attenuation and phase shift at each frequency in the series. The trouble is that there can potentially be a many frequencies and it is going to get boring. What if there were a graph that would provide us what we need at each frequency? There is. It is called a *Bode plot*.

4.4 Bode Plots

4.4.1 The Unit of Decibels

Before launching into this section, we will consider the unit of decibels, which is widely used in engineering. The amplitude of a frequency domain function $A(\omega)$ can be written in decibels by

$$A_{db} = 20\log_{10}|A(\omega)|. \tag{4.25}$$

The phase of $A(\omega)$ is expressed the same as before.

Example 4.32 Convert to decibels (dB):

a. $|A(\omega)| = 1$
b. $|A(\omega)| = 10$
c. $|A(\omega)| = 100$
d. $|A(\omega)| = 0.01$

Solution

a. $A_{db} = 20\log_{10}|A(\omega)| = 20\log_{10}(1) = 0$
b. $A_{db} = 20\log_{10}(10) = 20 \cdot 1 = 20$
c. $A_{db} = 20\log_{10}(10^2) = 40 \cdot 1 = 40$
d. $A_{db} = 20\log_{10}(10^{-2}) = -40 \cdot 1 = -40$

4.4.2 Introduction to Bode Plots

Suppose we have a system with the transfer function

$$H(s) = \frac{10^3}{s + 10^3} \tag{4.26}$$

This is a one-pole, low pass filter. We know that for any frequency, we are most interested in the magnitude and phase. However, instead of simply magnitude, it will be advantageous to look at the answer in decibels, which is

$$|H(\omega)|_{dB} = 20\log_{10}|H(\omega)|.$$

What will that give us? When $\omega \ll 10^3$,

$$H(\omega) = \frac{10^3}{j\omega + 10^3} \cong \frac{10^3}{10^3} = 1.$$

The logarithm of one is zero. What happens when omega is much greater than 10^3?

$$H(\omega) = \frac{10^3}{j\omega + 10^3} \cong \frac{10^3}{j\omega}.$$

and

$$|H(\omega)|_{dB} = 20\log_{10}\left|\frac{10^3}{j\omega}\right| = -20\log_{10}\left|\frac{j\omega}{10^3}\right|,$$

e.g., at $\omega = 10^4$

$$|H(\omega)|_{dB} = -20\log_{10}\left|\frac{10^4}{10^3}\right| = -20,$$

$\omega = 10^5$

$$|H(\omega)|_{dB} = -20\log_{10}\left|\frac{10^5}{10^3}\right| = -40,$$

i.e., it drops at the rate of 20 dB per decade. What about at $\omega = 10^3$?

$$|H(\omega)|_{dB} = -20\log_{10}\left|\frac{10^3}{j10^3 + 10^3}\right|$$

$$= -20\log_{10}\left|\frac{1}{\sqrt{2}}\right| = -3 \ dB.$$

Now consider the phase. We calculate this by

$$\angle H(\omega) = \frac{\angle 10^3}{\angle(j\omega + 10^3)} = -\angle(j\omega + 10^3),$$

when $\omega \ll 10^3$, $\angle H(j\omega) \cong 0$. When $\omega \gg 10^3$, $\angle H(j\omega) = -\angle(j\omega) = -90°$. When the frequency is not at one of these extremes, we have to calculate the phase.

The following MATLAB program graphs a Bode plot.:

```
% My_bode.m

w  = (10^1:2:10^5);
WW = log10(w);

% A one-pole LPR
num = [1e3];
den = [1 1e3];
```

```
g = tf(num,den)
H = freqs(num,den,w);

% Bode plots

subplot(2,1,1)
semilogx(w,20*log10(abs(H)),'k')
axis ( [ 10^2 10^5 -40 5 ])
set(gca,'fontsize',14)
ylabel('|H| db')
grid on
title('My-bode')

subplot(2,1,2)
semilogx(w,(180/pi)*angle(H),'k');
axis ( [ 10^1 10^5  -100 10 ])
set(gca,'YTick',[ -180 -90 -45 0   45 90 ])
set(gca,'fontsize',14)
xlabel('\omega (rad/sec)');
ylabel('/_ H (degrees)')
grid on
```

The results of this program are shown in Fig. 4.13.

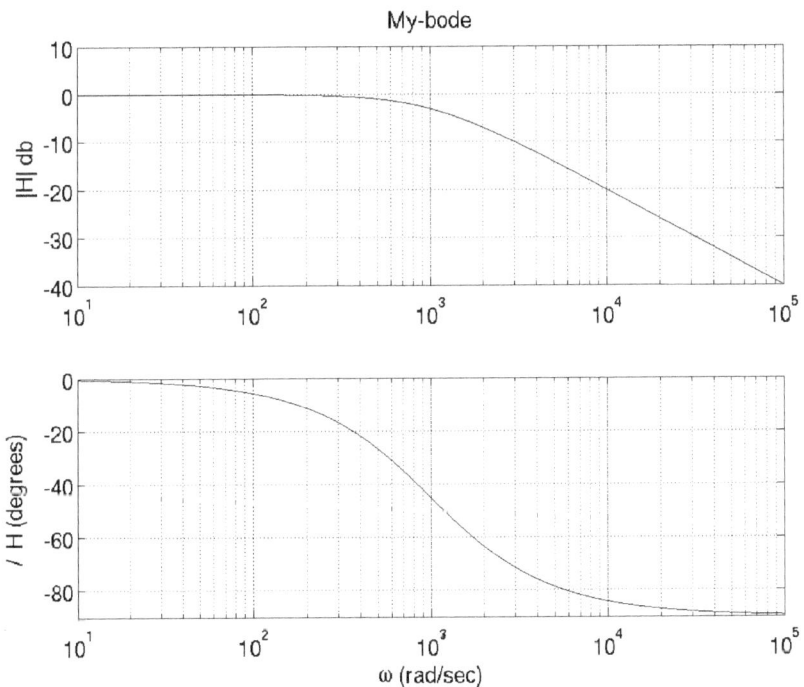

Figure 4.13. Bode plots of the one-pole low pass filter of Eq. (4.26).

If we were constructing a Bode plot of Eq. 4.26 by hand, we would first draw the asymptotes, i.e, 0 between $0 < \omega < 10^3$, and minus 20 dB per decade for $\omega > 10^3$. At $\omega = 10^3$, it must be -3 dB. Then connect the lines smoothly. To draw the phase, it is zero for $\omega \ll 10^3$ and 90 degrees for $\omega \gg 10^3$. At $\omega = 10^3$, it is 45 degrees. So draw the curve that shows this.

Now suppose the transfer function were

$$H(s) = \frac{10^6}{(s+10^3)^2}. \qquad (4.27)$$

This is clearly a two pole low pass filter. We could go through the same procedure, and find that above $\omega = 10^3$, the magnitude decreased at 40 dB per decade, and at $\omega = 10^3$ it is -6 dB. The phase ranges from 0 to -180 degrees; and right at $\omega = 10^3$ it is -90 degrees. This is shown in Fig. 4.14

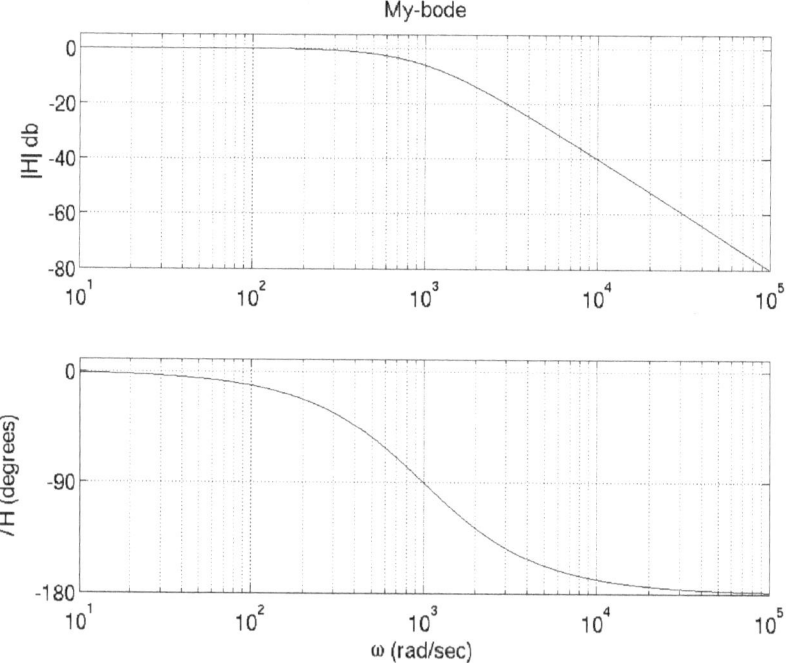

Figure 4.14. Bode plot of the two-pole low pass filter in Eq. (4.27).

Example 4.33

This example demonstrates one of the primary reasons for the Bode plots. Suppose the input to the two-pole low pass filter of Eq. (4.27) is
$$x(t) = 2\cos(2 \times 10^3 t + 30^\circ).$$
What is the result?

Solution

To determine this, we need only go to the Bode plot and find the corresponding frequency of $\omega = 2000$ rad/sec, as shown in Fig. 4.15. The magnitude is at about -14 dB. This corresponds to
$$-14 = 20 \log_{10} |H(\omega)|$$
or
$$|H(\omega)| = 10^{-14/20} = 0.196$$
Similarly, we read a phase shift of -125 degrees. So we would say the output is
$$y(t) = (.196) 2 \cos(2 \times 10^3 t + 30^\circ - 125^\circ)$$
$$= 0.392 \cos(2 \times 10^3 t - 95^\circ)$$

Check
$$H(\omega) = \frac{10^6}{(j2 \times 10^3 + 10^3)^2} = \frac{1}{(j2+1)^2}$$
$$= \frac{1}{(\sqrt{5} \angle 63^\circ)^2} = 0.2 \angle -126^\circ$$

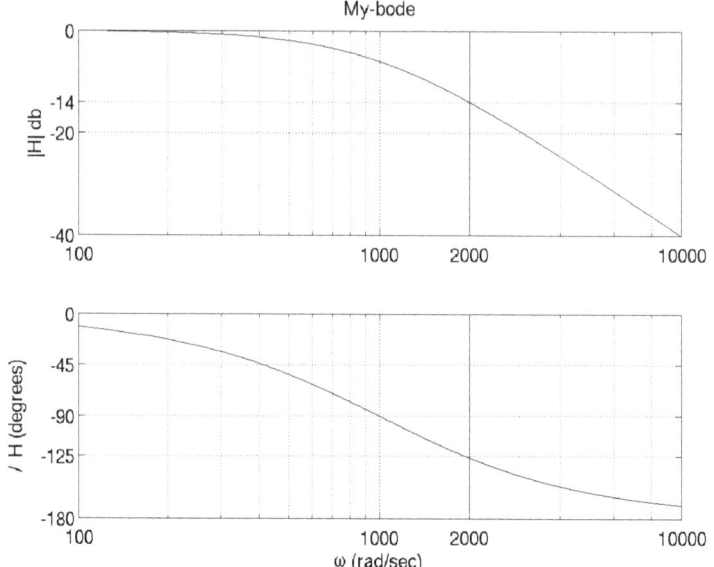

Figure 4.15. Using the Bode plot to determine attenuation and phase shift of Eq. (4.27) at $\omega = 2 \times 10^3$ rad/sec.

Look at the following transfer function:

$$H(s) = \frac{s + 10^3}{10^3}. \tag{4.28}$$

The zero at $\omega = 10^3$ means the magnitude will start rising at a rate of 20 dB/decade. Note that above $\omega = 10^3$, it looks like a *differentiator*.

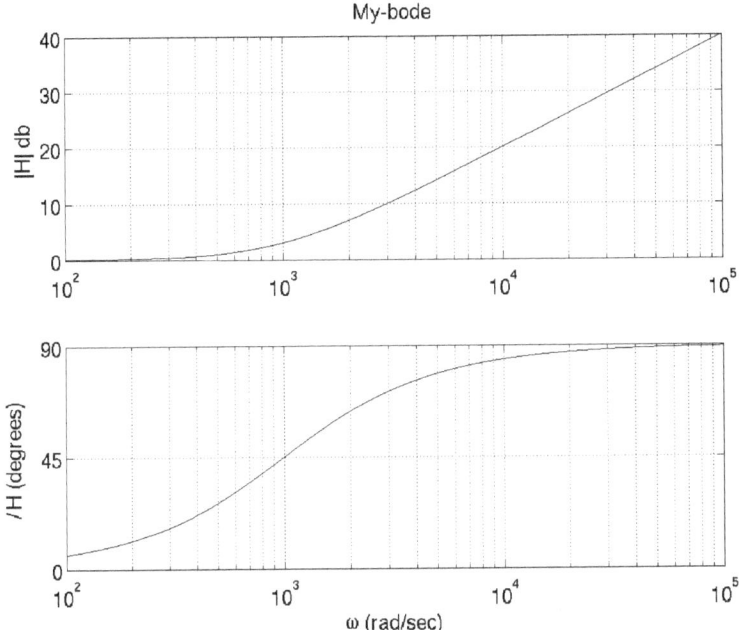

Figure 4.16. Bode plot of Eq. (4.28).

Now let us look at the filter described by

$$H(s) = 10^3 \frac{s}{(s+10)(s+10^3)}. \tag{4.29}$$

Equation (4.29) has a "zero," the s term in the numerator. When $\omega \ll 10$, the transfer function looks like this:

$$H(\omega)_{\omega \ll 10} = \frac{10^3 j\omega}{(10)(10^3)} = 10^{-1} j\omega.$$

This function looks like a differentiator for low frequencies. Remember that for both Laplace and Fourier transforms, $sX(s)$ or $j\omega X(j\omega)$ means "the Fourier (Laplace) transform of the derivative of $x(t)$." It looks like a differentiator until it approaches the first pole at $\omega = 10$. To find out what happens, let us see what it is at $\omega = 100$:

$$H(\omega)_{\omega=100} = 10^3 \frac{j100}{(j100+10)(j100+10^3)} \cong 10^3 \frac{j100}{(j100)(10^3)} = 1.$$

In other words, it has little frequency dependence, at least until the frequency starts approaching $\omega = 1000$. Above $\omega = 1000$, what happens?

$$H(\omega)_{\omega \Box 1000} = 10^3 \frac{j\omega}{(j\omega+10)(j\omega+10^3)} \cong \frac{10^3}{j\omega}.$$

Now we would say the function has started to look like an integrator. The Bode plot is given in Fig. 4.17.

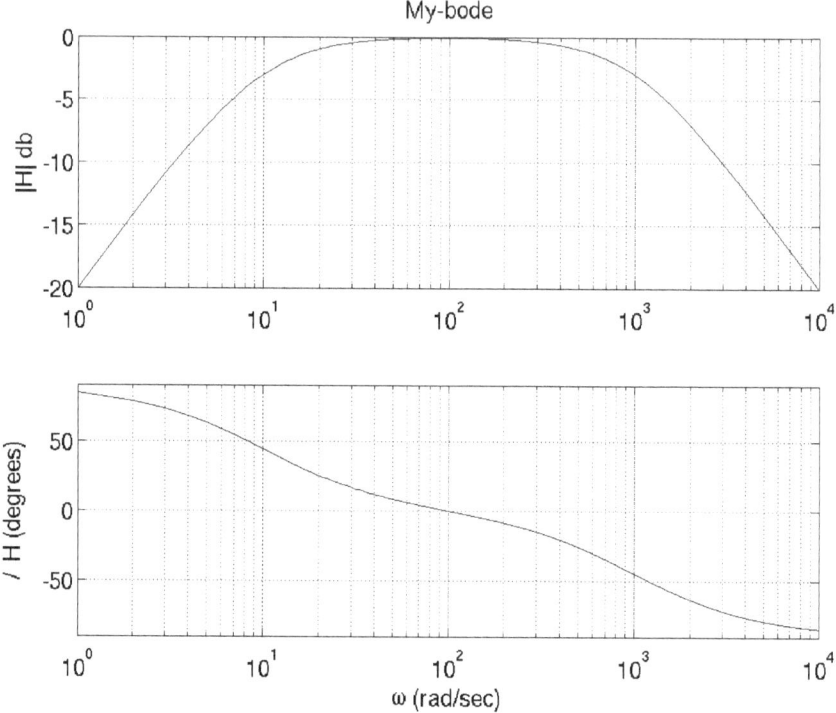

Figure 4.17. Bode plot of the band pass filter in Eq. (4.29).

Example 4.34 Suppose the series in Fig. 4.18, $T = 10$ ms, is put through the filter described by Eq. (4.29) and described by the Bode plot in Fig. 4.17. How many terms are needed to describe the output?

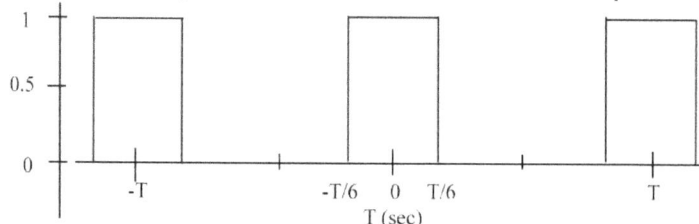

Figure 4.18. A series of pulse functions

Solution

We found earlier that the Fourier series is

$$x(t) = \sum_{n=-\infty}^{\infty} X_n e^{jn\omega_0 t} = \frac{1}{3} + \sum_{n=1}^{\infty} 2X_n \cos(n\omega_0 t)$$

$$X_n = \frac{\sin(\pi n/3)}{\pi n}$$

For $T = 10^{-2} s$

$$\omega_0 = \frac{2\pi}{T} = 2\pi \times 10^2 = 6.28 \times 10^2$$

$$2\omega_0 = 1.25 \times 10^3$$
$$3\omega_0 = 1.88 \times 10^3$$
$$4\omega_0 = 2.5 \times 10^3$$
$$5\omega_0 = 3.77 \times 10^3$$

To determine how much each frequency component is attenuated, we look at Fig. 4.17.

Clearly the DC term will not pass through. At $n = 1$, the filter attenuates about 2 dB or

$$20 \log_{10}|H| = -2$$
$$\log_{10}|H| = -0.1$$
$$|H_1| = 0.79$$

At n= 5, the filter amplitude is about -11 dB,

$$\log_{10}|H_5| = 0.28$$

Note also, that the magnitudes of the input terms are attenuated as the frequency increases. By n=5, the ratio of amplitudes between the n=1 term and the n= 5 term is

$$\frac{X_5}{X_1}\frac{|H_5|}{|H_1|} = \frac{1/5}{1/1}\frac{0.28}{0.79} = 0.07.$$

We will probably not be making a significant error by only using the first five terms. Actually, $X_3 = 0$.

For a calculation like the one above, it is often easier to rescale the Bode plot to cover only the area of interest, in this case between 500 rad/s and 5000 rad/s. Furthermore, it would be advantageous to just plot the magnitude instead of decibels. Reading the needed amplitude and phase values can be aided by using the XTick feature of MATLAB the find the values at the needed frequencies (Fig. 4.19).

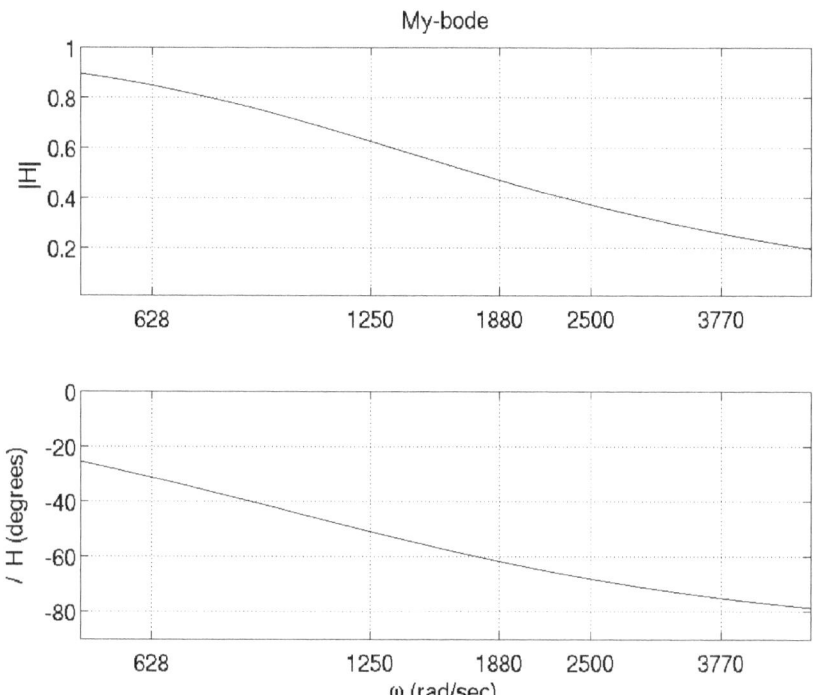

Figure 4.19. This is the same Bode plot as Fig 4.17, but the scale has been restricted to the frequencies of interest. Also, amplitude is plotted instead of decibels. The XTick feature of MATLAB is used by make reading parameters at the frequencies of interest easier.

4.4.3 Drawing Bode Plots

In drawing the magnitude, the following rules are used in first drawing the asymptotes:

1. Each pole, i.e., a term like $\dfrac{1}{s + \omega_0}$, will result in a

negative slope of -20 dB/decade once the frequency is above ω_0. A term $\dfrac{1}{s}$ means the plot starts with a -20 dB/decade slope.

2. Each zero, i.e., a term like $s + \omega_0$, will result in a positive slope of 20 dB/decade once the frequency is above ω_0. The term s means the plot starts with a 20 dB/decade slope.

In drawing the phase plot, the following rules apply:

3. Each pole $\dfrac{1}{s+\omega_0}$ will result in a negative 90 degree phase shift one decade beyond ω_0. It has no effect a decade before ω_0, and at ω_0 it results in a minus 45 degree phase shift.

4. Each zero $s + \omega_0$ will result in a positive 90 degree phase shift one decade beyond ω_0. It has no effect a decade before ω_0, and right at ω_0 it results in a 45 degree phase shift.

Example 4.35. Draw a Bode plot for the following transfer function:

$$H(s) = 10^{-4} \frac{s(s+10^4)}{(s+10^2)}$$

Solution

We see that there are zeros at $s = 0$ and $s = 10^4$ and a pole at $s = 10^2$. The zero at $s = 0$ means the plot is originally going up at 20 dB/decade. Then at $s = 10^2$ it flattens, and at $s = 10^4$ it starts up at 20 dB/decade again. To get a reference point, we evaluate at $s = 10^3$

$$H(10^3) \cong 10^{-4} \left. \frac{s(10^4)}{s} \right|_{s=10^3} = 1.$$

Furthermore, we see the angle

$$\angle H(10^3) \cong 10^{-4} \left. \frac{s(10^4)}{s} \right|_{s=10^3} = \arctan\left(\frac{j\omega}{j\omega}\right) = 0^o.$$

From there we can draw the asymptotes, as shown in Fig. 4.19.

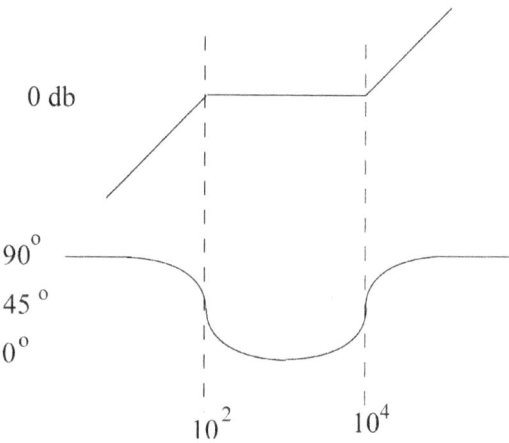

Figure 4.20. Asymptotic approximation of the bode plot for

$$H(s) = 10^{-4} \frac{s(s+10^4)}{(s+10^2)}.$$

The MATLAB commands

```
num = 1e2*[1 1e2];
den = [1 1e4 0]

g = tf(num,den)
H = freqs(num,den,w);
```

in the program **my_bode.m** produce Fig. 4.21.

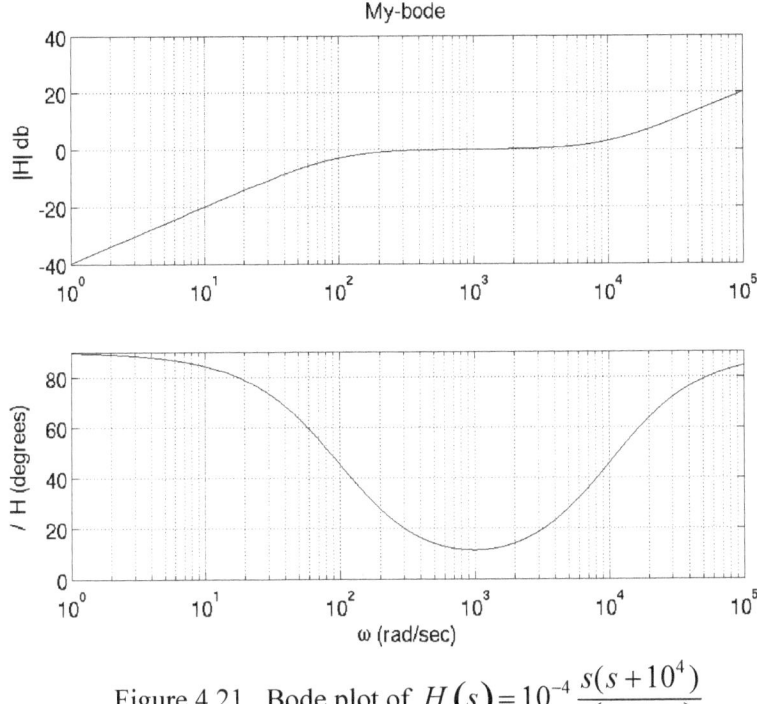

Figure 4.21. Bode plot of $H(s) = 10^{-4} \dfrac{s(s+10^4)}{(s+10^2)}$

Example 4.36. Draw the Bode plot of the transfer function of the following circuit assuming the following values: L = 10 mH, C = 1 μF, and R = 25 Ω.

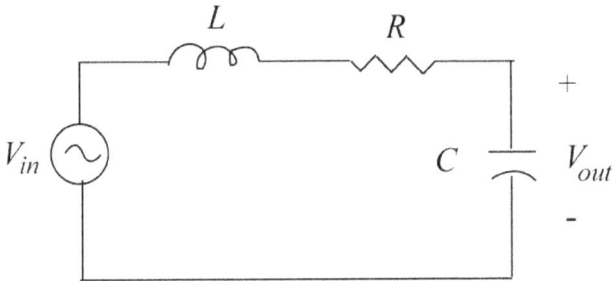

Figure 4.22. An RLC circuit.

Solution

$$H(s) = \frac{V_{out}(s)}{V_{in}(s)} = \frac{1/sC}{sL + R + 1/sC} = \frac{1/LC}{s^2 + s(R/L) + 1/LC}$$

$$= \frac{10^8}{s^2 + s(2.5 \times 10^3) + 10^8}$$

We see that the denominator is second order, so we will look for the position of the two poles

$$p = \frac{-2.5 \times 10^3 \pm \sqrt{(2.5 \times 10^3)^2 - 4 \cdot 10^8}}{2} = (-1.25 \pm j9.92) \times 10^3.$$

From this we see that the corresponding corner frequency is $\omega = 9.92 \ kHz$. What is $H(s)$ at this value? For simplicity, we will use $s = j1000$.

$$H(s) = \frac{10^8}{(j10^4)^2 + (2.5 \times 10^3)(j10^4) + 10^8}$$

$$= \frac{10^8}{-10^8 + j(2.5 \times 10^7) + 10^8} = \frac{4}{j} = 4\angle -90°,$$

$$20\log_{10}(4) = 12 \ db.$$

Instead of the -6 dB attenuation we expect at the corner frequency of a two pole low pass filter, there is a *gain*. This gain is known as *overshoot*. At the resonant frequency of the circuit, this value of R is not enough to completely dampen the resonance. The Bode plot is given below.

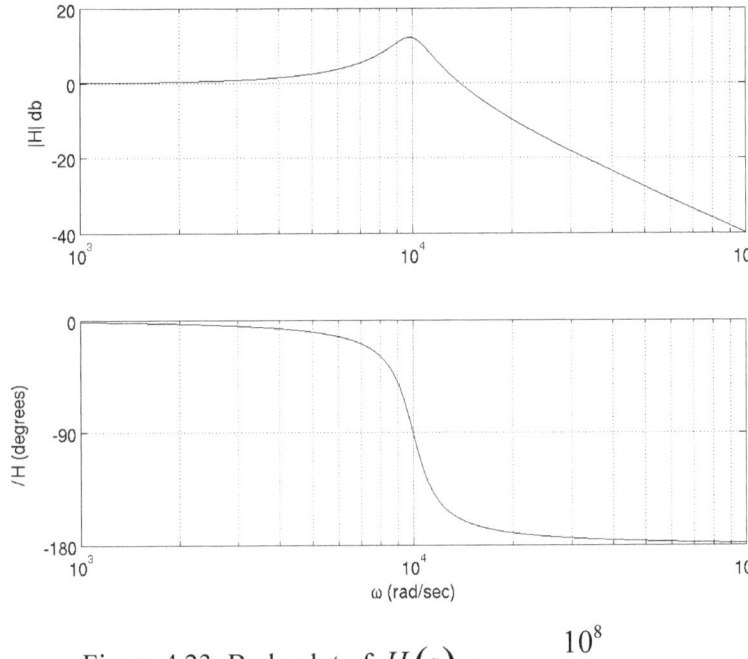

Figure 4.23. Bode plot of $H(s) = \dfrac{10^8}{s^2 + 2500s + 10^8}$.

4.4.4. Determining the Transfer Function from a Bode Plot.

The Bode plot can be thought of as a detailed, quantitative description of a function, usually a transfer function. As we have seen, if we have a Bode plot describing the transfer function we know how a signal at any frequency will be effected. In this section, we will describe how to construct the transfer function from a Bode plot. This is often useful in determining the qualitative features of a transfer function.

To determine a transfer function from a Bode plot, first write down the poles and zeros by observing the slopes on the magnitude plot. Then from a flat point on the slope, evaluate the magnitude. This will give the multiplying constant.

Sometimes, plots created from a transfer function with multiple poles and zeros can be evaluated with the help of the phase plot. For instance, if the plot ends at -90 degrees, there was one more pole than zeros. If it starts at 90 degrees, it means that there is a term $1/s$.

Example 4.37 Find the transfer function corresponding to the Bode plot in Fig. 4.24.

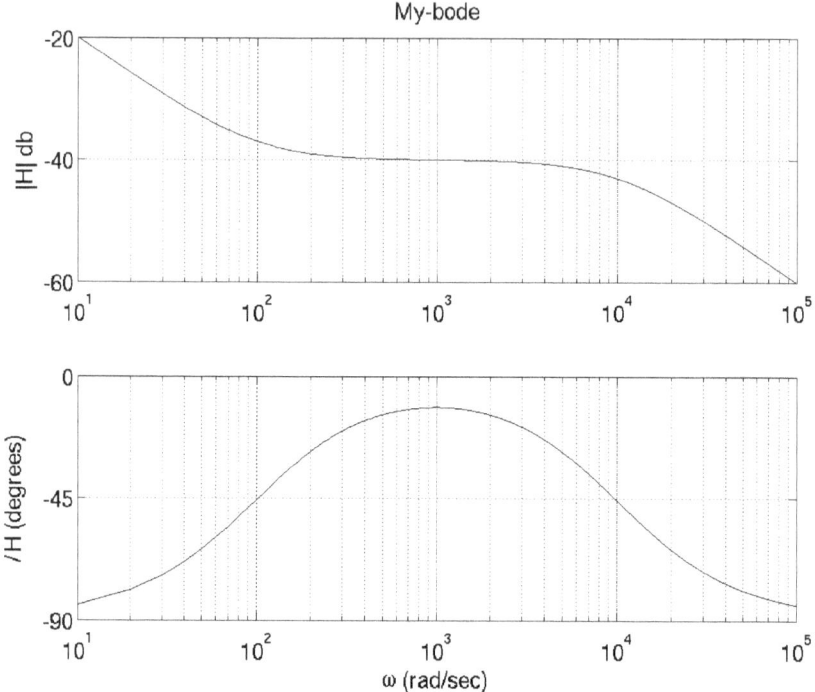

Figure 4.24. Bode plot of a transfer function.

Solution

It this situation, more can be learned by looking at the angle. For the very low frequencies, the angle is -90 degrees, so there must be a factor 1/s. Then at $\omega = 10^2$ the phase passes through 45 degrees on its way up, so there is a zero at $\omega = 10^2$, giving a factor $s + 10^2$ in the numerator. Then it passes through -45 degrees on its way back down to -90 degrees at $\omega = 10^4$. So far, we suspect our transfer function looks something like

$$H(s) = K \frac{s + 10^2}{s(s + 10^4)},$$

where K is an unknown constant. To get K, we have to look at the magnitude. At $\omega = 10^3$ it appears to be -40 dB, so

$$-40db = 20\log_{10}\left[K\frac{j\omega+10^2}{j\omega(j\omega+10^4)}\right]_{\omega=10^3}$$

$$-2 \cong \log_{10}\left[K\frac{j10^3}{j10^3(10^4)}\right] = \log_{10}\left[K10^{-4}\right]$$

$$= \log_{10} K - 4$$

$$K = 10^2$$

$$H(s) = 10^2 \frac{s+10^2}{s(s+10^4)}$$

4.5 Block Diagrams in the Frequency Domain

Everything we previously learned in Chapter 3 about block diagrams for Laplace transforms holds true for Fourier transforms. Only the applications change.

Example 4.38 Suppose the periodic signal of Fig. 4.25 (a) is put into a system described by the block diagram in Fig. 4.25 (b). What is the output $y(t)$?

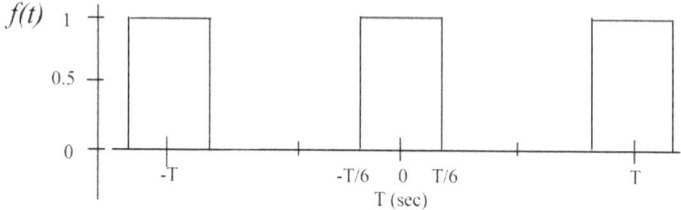

(a) A periodic signal; T = 1 sec.

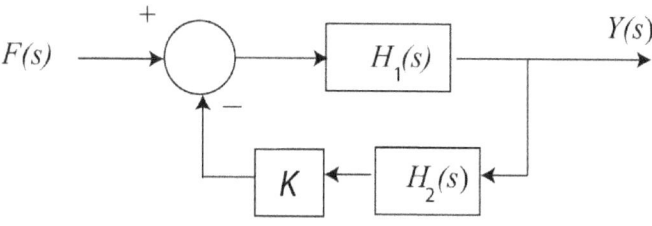

(b) Block diagram

Figure 4.25. (a) A periodic signal $f(t)$, which is the input to a system described by the block diagram (b).

$$H_1(s) = \frac{1}{s+3}$$

$$H_2(s) = \frac{1}{s} \quad K = 1$$

Solution

First, we need an expression for $f(t)$. Using the complex series

$$X_n = \frac{1}{T}\int_{-T/2}^{T/2} p_{T/3}(t)e^{-jn\omega_0 t}dt = \frac{1}{T}\left\{\frac{T}{3}\frac{\sin(\omega T/6)}{\omega T/6}\right\}_{\omega=n\omega_0}$$

$$= \frac{1}{3}\frac{\sin(2\pi n/6)}{2\pi n/6} = \frac{1}{3}\frac{\sin(\pi n/3)}{\pi n/3}$$

Because we see that $X_{-n} = X_n$,

$$f(t) = \sum_{n=-\infty}^{\infty} X_n e^{jn\omega_0 t} = \frac{1}{3} + \sum_{n=1}^{\infty} 2X_n \cos(n\omega_0 t).$$

The diagram has the transfer function

$$H(s) = \frac{\frac{1}{s+3}}{1+\frac{1}{s}\frac{1}{s+3}} = \frac{s}{s^2+3s+1}$$

In Chapter 3, we used Laplace transforms because we were using causal functions. Now we are dealing with sinusoidal functions, so it is more appropriate to use frequency domain functions. We can switch the above transfer functions from the s domain to the frequency domain by the replacing s with $j\omega$

$$H(s)\big|_{s=j\omega} = H(\omega) = \frac{j\omega}{(1-\omega^2)+3j\omega}.$$

We know that we are only going to be interested in the frequencies of the series, i.e., $\omega = n\omega_0$, so

$$H(n\omega_0) = \frac{jn\omega_0}{(1-n^2\omega_0^2)+3jn\omega_0}$$

So the output of the system will be

$$y(t) = \frac{1}{3}|H(0)| + \sum_{n=1}^{\infty} 2|H(n\omega_0)|X_n \cos[n\omega_0 t + \angle H(n\omega_0)]$$

Since $T=1$, $\omega_0 = 2\pi$

$$H(n\omega_0) = \frac{j2\pi n}{(1-n^2(2\pi)^2)+6\pi jn}.$$

When $n=0$,

$$H(0) = 0,$$

so no DC term gets through.

$$H(\omega_0) = \frac{j2\pi}{(1-(2\pi)^2)+6\pi j} = 0.1466\angle -64°$$

$$H(2\omega_0) = \frac{j4\pi}{(1-4(2\pi)^2)+12\pi j} = 0.078\angle -77°$$

The output to the block diagram is:

$$y(t) = \sum_{n=1}^{\infty} 2X_n |H(n\omega_0)| \cos[n\omega_0 t + \angle H(n\omega_0)].$$

Since $X_3 = 0$, there is no reason to calculate the $H(3\omega_0)$. From $n=4$ forward, the magnitudes of both $H(n\omega_0)$ and X_n drop as n increases. We will use the first two terms to approximate the output as

$$y(t) \cong 2\frac{\sin(\pi/3)}{\pi}(0.1466)\cos(2\pi t - 64°)$$
$$+ 2\frac{\sin(\pi/6)}{2\pi}(0.078)\cos(4\pi t - 7°)$$
$$= 0.080\cos(2\pi t - 64°) + 0.012\cos(4\pi t - 7°).$$

Of course, we can expedite the process by using a Bode plot displaying amplitude and phase over the frequencies of interest, as shown in Fig. 4.26.

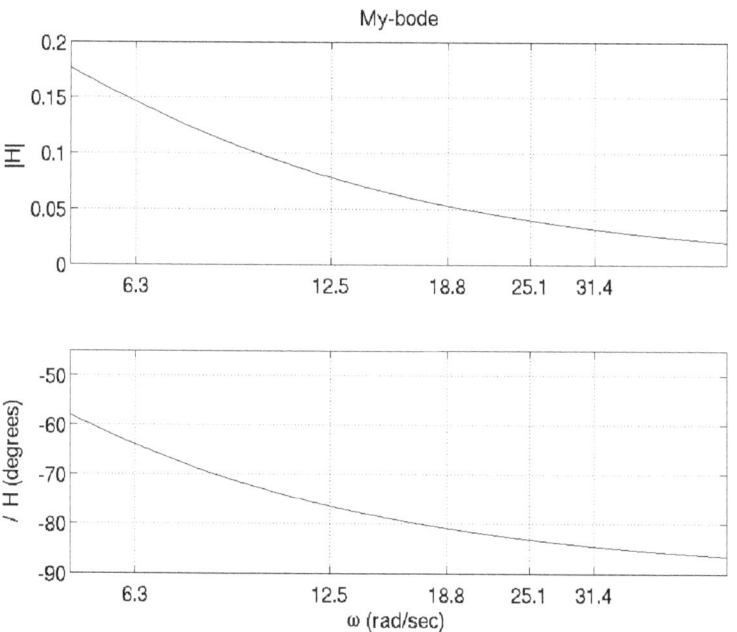

Figure 4.26. Bode plot of the transfer function over the region of interest.

Table 4.1 Some Common Fourier Transform Pairs

$x(t)$	$X(\omega)$
$\delta(t)$	1
1	$2\pi\delta(\omega)$
$e^{j\omega_0 t}$	$2\pi\delta(\omega-\omega_0)$
$\operatorname{sgn}(t)$	$\dfrac{2}{j\omega}$
$u_h(t)$	$\dfrac{1}{j\omega}+\pi\delta(\omega)$
$e^{-\alpha t}u_h(t),\ \alpha>0$	$\dfrac{1}{j\omega+\alpha}$
$\cos(\omega_0 t)$	$\pi\left[\delta(\omega+\omega_0)+\delta(\omega-\omega_0)\right]$
$\sin(\omega_0 t)$	$j\pi\left[\delta(\omega+\omega_0)-\delta(\omega-\omega_0)\right]$
$e^{-\alpha|t|},\ \alpha>0$	$\dfrac{2\alpha}{\alpha^2+\omega^2}$
$\dfrac{1}{\sqrt{2\pi}\sigma}e^{-\frac{1}{2}\left(\frac{t}{\sigma}\right)^2}$	$e^{-\sigma^2\omega^2/2}$
$p_\tau^h(t)$	$\tau\dfrac{\sin(\omega\tau/2)}{\omega\tau/2}=\tau\operatorname{sinc}(\omega\tau/2)$
$\Delta_\tau(t)$	$\dfrac{\tau}{2}\operatorname{sinc}^2\left(\dfrac{\omega\tau}{4}\right)$

Table 4.2 Some Properties of Fourier Transforms.

	$x(t)$	$X(s)$
1.	$\alpha x_1(t) + \beta x_2(t)$	$\alpha X_1(\omega) + \beta X_2(\omega)$
2.	$x(t-t_0)$	$e^{-j\omega t_0} X(\omega)$
3.	$\dfrac{d^n x(t)}{dt^n}$	$(j\omega)^n X(\omega)$
4.	$x_1(t) * x_2(t)$	$X_1(\omega) \cdot X_2(\omega)$
5.	$\int_{-\infty}^{t} x(\tau)d\tau$	$\dfrac{1}{j\omega} X(\omega) + \pi X(0)\delta(\omega)$
6.	$t^n x(t)$	$(j)^n \dfrac{d^n}{d\omega^n} X(\omega)$
7.	$x_1(t) \cdot x_2(t)$	$\dfrac{1}{2\pi} X_1(\omega) * X_2(\omega)$
8.	$x(-t)$	$X(-\omega)$
9.a	$x(t)e^{j\omega_0 t}$	$X(\omega - \omega_0)$
9.b	$x(t) \cdot \cos(\omega_1 t)$	$\dfrac{1}{2}\left[X(\omega + \omega_1) + X(\omega - \omega_1)\right]$
9.c	$x(t) \cdot \sin(\omega_1 t)$	$\dfrac{j}{2}\left[X(\omega + \omega_1) - X(\omega - \omega_1)\right]$
10.a	$X(t)$	$2\pi x(-\omega)$
10.b	$X(-t)$	$2\pi x(\omega)$
11.	Parseval's Theorem:	$\displaystyle\int_{-\infty}^{\infty} |x(t)|^2 \, dt = \dfrac{1}{2\pi} \int_{-\infty}^{\infty} |X(\omega)|^2 \, d\omega$

References

1. Z. Gajic, *Linear Dynamic Systems and Signals,* Upper Saddle River, NJ: Prentice Hall, 2003.

2. R. Bracewell, *The Fourier Transforms and Its Applications,* New York, NY: McGraw-Hill, 1999.

3. A. Papoulis, *The Fourier Integral and Its Applications,* New York, NY: McGraw-Hill, 1962.

3. E. O. Brigham, *The Fast Fourier Transform and Its Applications,* Englewood Cliffs, NJ: Prentice Hall, 1988.

3. M. J. Roberts, *Signals and Systems—Analysis Using Transform Methods and MATLAB,* 2nd ed, New, York, NY: McGraw Hill, 2012.

Problems

4.1.1 The impulse response of a system is given by
$$h(t) = 10^6 \left(e^{-t/\tau_1} - e^{-t/\tau_2} \right) u(t) \quad \tau_1 = 1\ \mu s, \quad \tau_3 = 3\ \mu s.$$
a). What is the response to $x_{in}(t) = 10\cos(\omega_0 t)$ $\omega_0 = 500\ kHz$?
b). What is the response to $x_{in}(t) = 10\cos(\omega_0 t)$ $\omega_0 = 10\ kHz$?

4.1.2. A linear system has the impulse response
$$h(t) = u(t) - u(t - 0.001).$$
What is the transfer function in the Fourier domain?

4.1.3. Find the Fourier transform of
$$x(t) = e^{-t/0.005} \cos(10^3 t) u(t)$$

4.1.4 A half-cosine function is defined by
$$x(t) = \cos(2\pi 10^3 t) \quad -5\ ms \le t \le 5\ ms.$$
Find the Fourier transform.

4.1.5. Find the Fourier transform of
$$f(t) = r(t) - 1.5r(t-1) + 0.5r(t-3)$$
Hint: Think integration property.

4.1.6. Calculate the following integral using Parseval's theorem
$$\int_0^\infty \frac{\sin^2(10^6 t)}{(10^6 t)^2} dt.$$

4.1.7 Find the Fourier transform of the signal
$$x(t) = e^{-(3t)^2}$$

4.1.8 Show that the Fourier transform of the triangle function $\Delta_\tau(t)$ is
$$\frac{\tau}{4}\operatorname{sinc}^2\left(\frac{\omega \tau}{4}\right).$$

4.1.9. Evaluate the following integral
$$I = \int_{-\infty}^{\infty} \left(\frac{2 \cdot 5}{5^2 + \omega^2}\right)^2 d\omega$$

4.1.10 Use Fourier transforms to evaluate the following integrals:

a). $$I = \int_0^\infty \exp\left(-\frac{1}{2}\left(\frac{t}{10^{-3}}\right)^2\right) \cos(10^4 t) dt$$

b). $$I_b = \int_0^\infty e^{-5t} \sin(10t) dt$$

4.2.1. Evaluate the following integral
$$\int_{-\infty}^{\infty} \Delta_4(t) \cos(3\pi t) dt$$

4.2.2 The diagram below represents one period of a time series that extends infinitely in each direction. Write the Fourier series of this signal. T= 1 second. Your answer should be a real series (i.e., not complex functions).

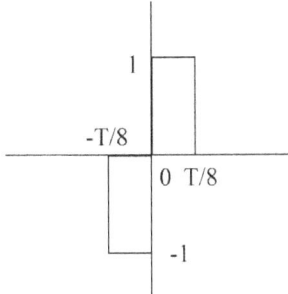

4.2.3. The pattern below extends infinitely in each direction. The interval is one second. Write a Fourier series. Your final answer should be a sine series.

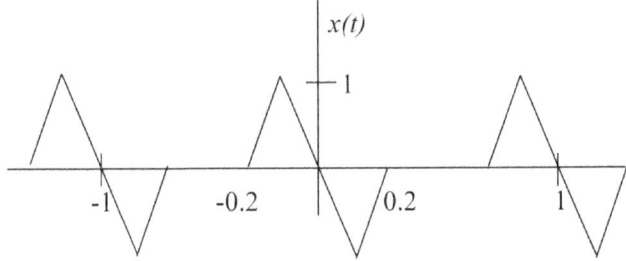

4.2.4. Write a Fourier series to describe the function below. You may assume that it extends infinitely in both directions. The amplitude of the delta functions is one. The time scale is seconds. Your final answer should be a sine and/or cosines series.

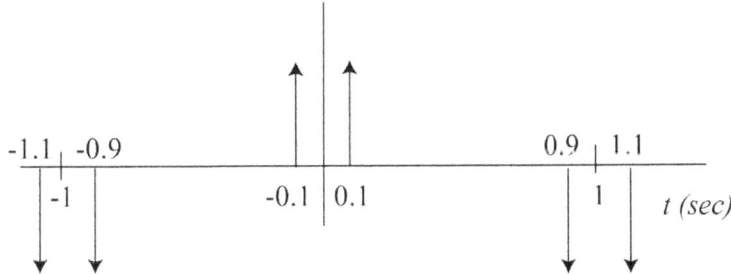

4.2.5 The series below is made up of function of the form
$$f(t) = e^{-50t^2}.$$
Write the Fourier series for T = 0.1 sec.

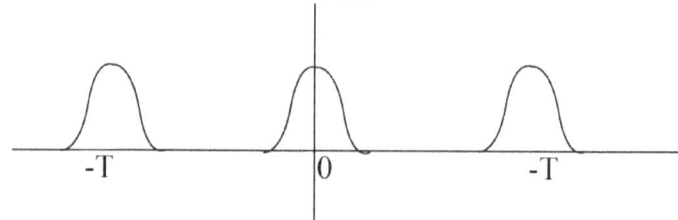

4.2.6 The following sequence extends infinitely in both directions. Write the Fourier series.

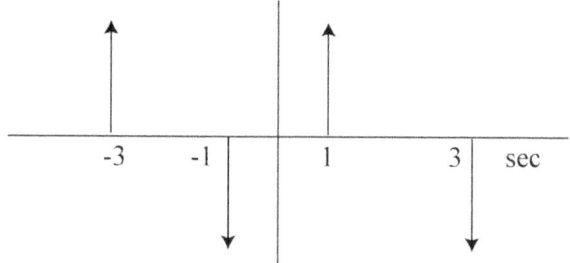

4.2.7 The following sequence extends infinitely in both directions. Write the Fourier series.

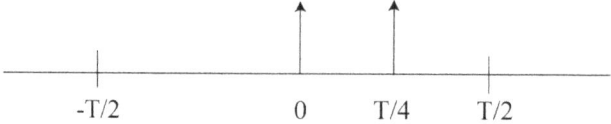

4.3.1 Solve for y(t):
$$\frac{d^2y(t)}{dt^2} + (2\times10^3)\frac{dy(t)}{dt} + 10^6 y(t) = 10^6 x(t)$$
$$x(t) = \sum_{n=1}^{\infty} \frac{1}{2n^2}\cos(10^3 nt)$$
Include enough terms for five percent accuracy.

4.3.2. The signal below extends infinitely in each direction. Assume the amplitude of the pulses is one, and that each pulse is 1 msec wide. Write a Fourier series for this signal.

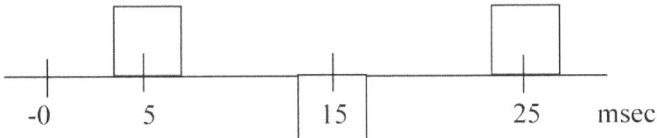

4.3.3. In the figure below, the two pulses repeat at one msec intervals. This series goes through a filter defined by

$$H(\omega) = \frac{10^4}{10^4 + j\omega}.$$

Write out the first three non-zero terms of the series that comes out of the filter. Assume the magnitude of the delta functions is 0.001.

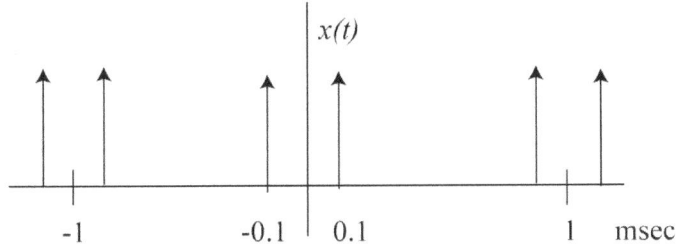

4.3.4. Each pulse in the series below has the form

$$x_0(t) = e^{-(t/\tau)^2} \quad \tau = 2\,\mu\text{sec} \quad T = 10\,\mu\text{sec}.$$

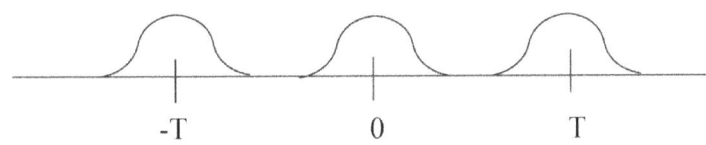

This series is the input to a system described by

$$\frac{dy(t)}{dt} + 2 \times 10^5 y(t) = 10^5 x(t).$$

What is y(t)?

4.4.1. Draw the Bode plot of

$$H(s) = \frac{10^6 (s + 10^2)}{(s^2 + 1.01 \times 10^6 + 10^{10})}$$

4.4.2. Repeat problem 4.3.1 using Bode plots

4.4.3 The pattern illustrated below repeats at 0.5 msec intervals. It is run through a filter in which the transfer function is described by the Bode plot. What comes out (to within ten percent accuracy)?

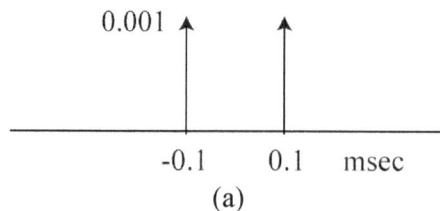

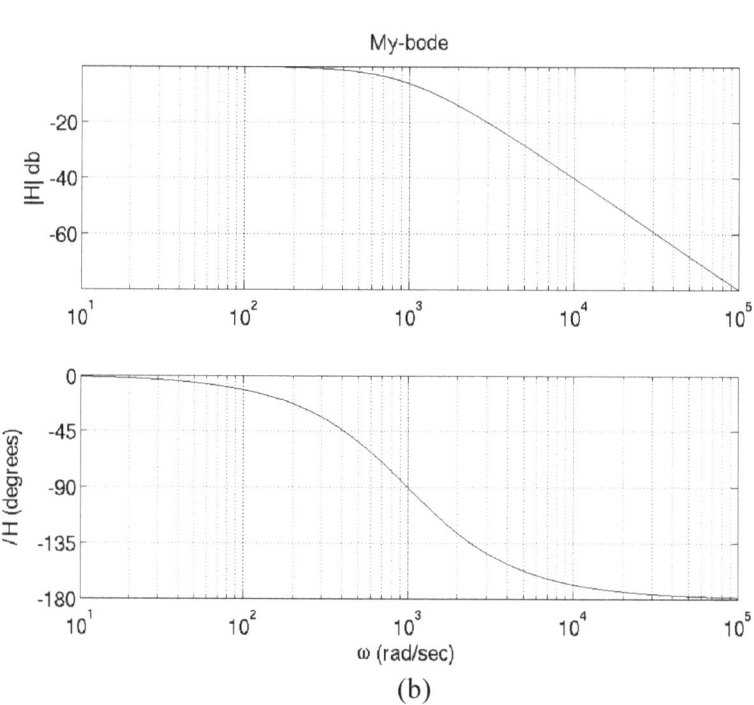

Figure 4.35. (a) One time period of a signal that repeats at 0.5 msec intervals; (b) the Bode plot of a filter.

4.5.1 Find the inverse Fourier Transform of $G(\omega) = 5e^{-2|\omega|}$

4.5.2 Find the Fourier transform of $y(t) = \dfrac{1}{t}$

4.5.3 Write the Fourier series of the following periodic signal. Each pulse has the form
$$x(t) = \operatorname{sinc}^2(8\pi t) \qquad T = 1$$

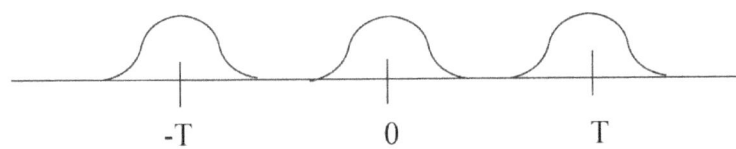

-T   0   T

A series of pulse at intervals of one second.

4.5.4. Find the inverse Fourier transform of
$$X(j\omega) = u(\omega) - u(\omega - 4)$$

4.5.5. Draw the Bode plot of
$$H(s) = \dfrac{10^3 s}{(s+10)(s+10^3)}$$

4.5.6 In Fig. 4.5.3, the input $x(t)$ is a series of pulses give by
$$x(t) = \sum_{n=-\infty}^{\infty} 2\pi \times 10^{-4} \delta(t - nT)$$

where $T = 2\pi \times 10^{-4}$ sec. The first transfer function is
$$H_1(\omega) = \dfrac{10^3}{j\omega + 10^3}.$$

The second transfer function H_2 is given by the Bode plot. Write the two most significant terms of the Fourier series of $y(t)$ to 10 percent accuracy.

A series block diagram.

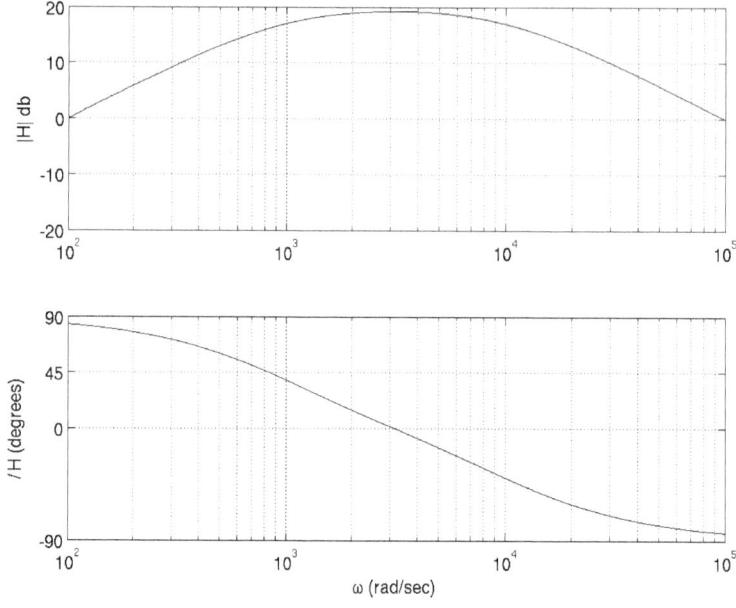

Bode plot of of H2.

Chapter 5. Z Transforms

Until now, we have been dealing with continuous signals. When we had causal signals, we used the Laplace transform. For non-causal signals, we used the Fourier transform. In this chapter, we will deal with discrete signals, i.e., signals that only have values at a specific interval. To deal with these discrete signals, we use the Z transform.

5.1 Discrete Signals

If we start with a continuous signal $f(t)$ and sample at constant intervals T, we get a discrete-time signal

$$f[k] = \sum_{k=0}^{\infty} f_k \delta(t - k \cdot T). \tag{5.1}$$

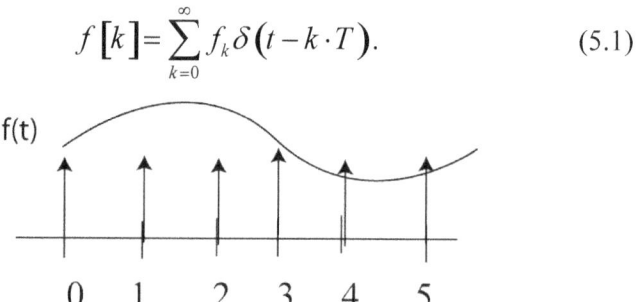

Figure 5.1. A signal $f(t)$ that is sampled at an interval T.

There are many discrete functions that are similar to continuous functions:

$$\delta[k] = \begin{cases} 1 & k = 0 \\ 0 & k \neq 0 \end{cases}$$

$$u[k] = \begin{cases} 1 & k \geq 0 \\ 0 & k < 0 \end{cases}$$

$$r[k] = \begin{cases} k & k \geq 0 \\ 0 & k < 0 \end{cases}$$

Previously we described physical systems using differential equations. Clearly, we can no longer do that. However, we can make use of the following approximation to convert continuous derivatives to difference equations:

$$\frac{dy(t)}{dt} \cong \frac{y(t+T)-y(t)}{T}. \tag{5.2}$$

Suppose we are looking to convert the following differential equation to a difference equation:

$$\frac{dy(t)}{dt}+\alpha y(t)= f(t). \tag{5.3}$$

Using Eq. (5.1) and (5.2) we can rewrite Eq. (5.3) as,

$$\frac{y[k+1]-y[k]}{T}+\alpha y[k]= f[k],$$

or

$$y[k+1]= y[k]-\alpha T y[k]+Tf[k]$$
$$= (1-\alpha T)y[k]+Tf[k].$$

As an example, look at the following difference equation

$$y[k]-0.5y[k-1]= x[k]. \tag{5.4}$$

Suppose the system is "at rest," i.e., with no initial conditions. Take the case where

$$x[k]= \delta[k].$$

Table 5.1 tracks the values of $y[k]$ using Eq. (5.4) rewritten in the following form:

$$y[k]= 0.5y[k-1]+ x[k].$$

Table 5.1.

k	y[k] y[k-1]	x[k]
0	1 0	1
1	0.5 1	0
2	0.25 0.5	0
3	0.125 0.25	0

We surmise that the output can be written

$$y[k]= h[k]= (0.5)^k u[k]. \tag{5.5}$$

Since this was the response to an impulse $\delta[k]$, we call it the impulse response, as we did with continuous systems. A graphs of $h[k]$ is shown in Fig. 5.2.

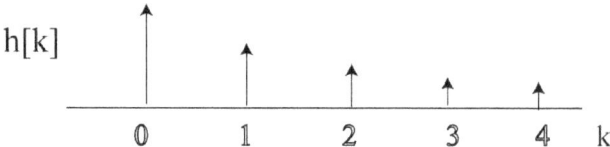

Figure 5.2. Graphs of the impulse response $h[k] = (0.5)^k u[k]$.

There are several discrete theorems that are similar to the continuous theorems:

1. The Sifting Theorem

$$\sum_{k=-\infty}^{\infty} f[k]\delta[k-k_0] = f[k_0] \qquad (5.6)$$

In discrete functions, summation takes the place of integration.

Notice that

$$\delta[k-k_0] = \begin{cases} 1 & k = k_0 \\ 0 & k \neq 0, \end{cases}$$

and therefore,

$$f[k]\delta[k-k_0] = f[k_0]\delta[k-k_0].$$

2. Convolution of Discrete-Time Signals

The convolution of two causal function, *h[k]* and *x[k]* is given by the following equation,

$$y[k] = \sum_{m=0}^{k} h[m]x[k-m]. \qquad (5.7)$$

Example 5.1. Determine the convolution of the two functions in Fig. 5.3.

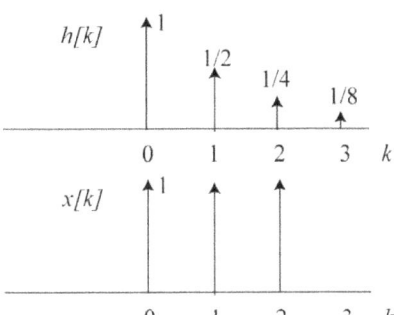

Figure 5.3. Two causal functions; *h[k]* extends infinitely but *x[k]* has only three terms.

Solution

To see the convolution graphically, we must flip one of the functions around the *k=0* axis. In Fig. 5.4 a, we have flipped *x[k]*. Using Eq. (5.7), we can calculate *y[0]* as shown. For increasing values of *k*, *x[k]* moves to the right, giving as increasing overlap with *h[k]*, as shown in Fig.5.4(b-d). The convolution continues to infinity, but the first few terms are shown in Fig. 5.5.

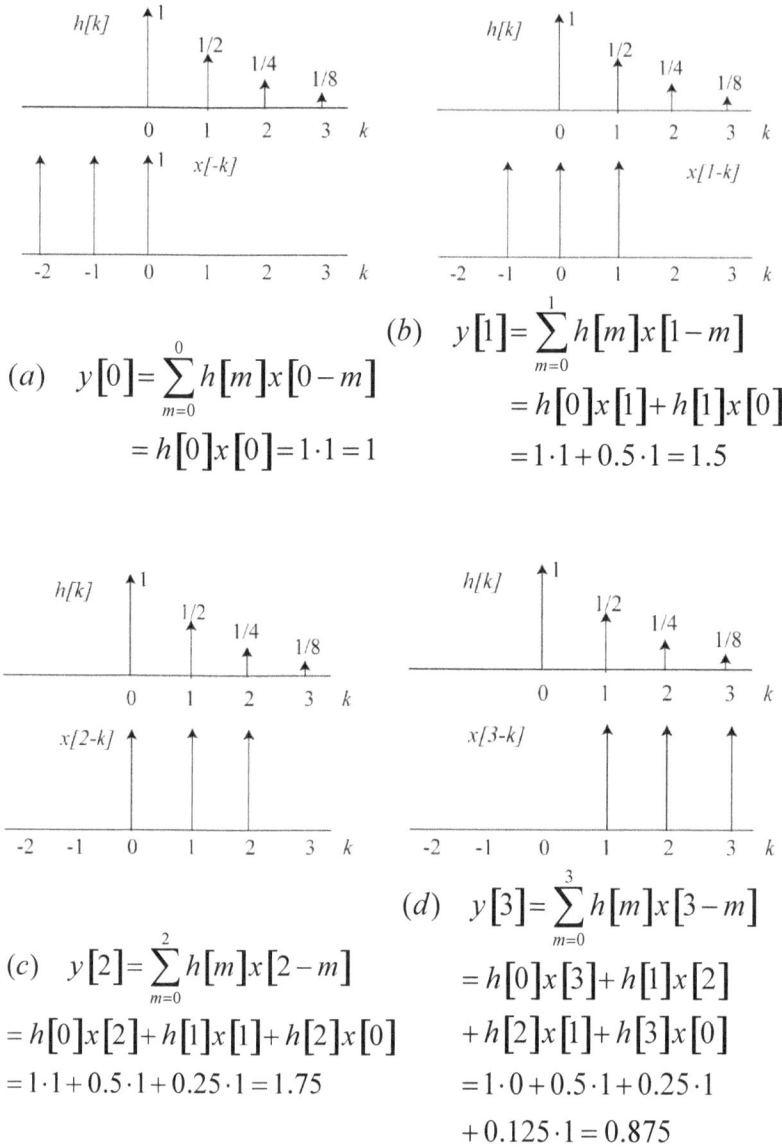

(a) $y[0] = \sum_{m=0}^{0} h[m]x[0-m]$
$= h[0]x[0] = 1 \cdot 1 = 1$

(b) $y[1] = \sum_{m=0}^{1} h[m]x[1-m]$
$= h[0]x[1] + h[1]x[0]$
$= 1 \cdot 1 + 0.5 \cdot 1 = 1.5$

(c) $y[2] = \sum_{m=0}^{2} h[m]x[2-m]$
$= h[0]x[2] + h[1]x[1] + h[2]x[0]$
$= 1 \cdot 1 + 0.5 \cdot 1 + 0.25 \cdot 1 = 1.75$

(d) $y[3] = \sum_{m=0}^{3} h[m]x[3-m]$
$= h[0]x[3] + h[1]x[2]$
$+ h[2]x[1] + h[3]x[0]$
$= 1 \cdot 0 + 0.5 \cdot 1 + 0.25 \cdot 1$
$+ 0.125 \cdot 1 = 0.875$

Figure 5.4. The first four iterations of the calculation of Eq. (5.7).

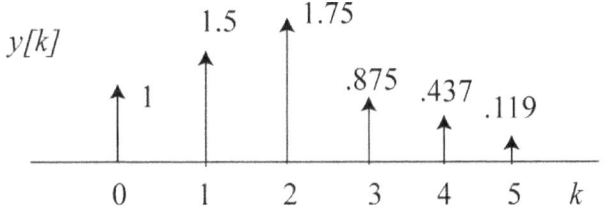
Figure 5.5. The convolution of h[k] and x[k] in Fig. 5.4.

The discrete convolution has many of the same properties as the corresponding continuous convolution:

i. Commutativity
$$f_1[k] * f_2[k] = f_2[k] * f_1[k]$$

ii. Distributivity
$$f_1[k] * \{f_2[k] + f_3[k]\} = f_1[k] * f_2[k] + f_1[k] * f_3[k]$$

iii. Associativity
$$f_1[k] * \{f_2[k] * f_3[k]\} = \{f_1[k] * f_{32}[k]\} * f_3[k]$$

The following property is unique to discrete convolution:

iv. Duration

If $f_1[k]$ is duration M_1,
and $f_2[k]$ is duration M_2,
then $f_1[k] * f_2[k]$ is duration $M_1 + M_2$.

Note: Duration is *not* the number of terms; it is the distance between the first and last terms.

v. Time-Shifting
$$f_1[k - k_1] * f_2[k] = f[k - k_1],$$
$$f_1[k - k_1] * f_2[k - k_2] = f[k - k_1 - k_2].$$

The convolution of the unit step with any causal signal produces a summation
$$f[k] * u[k] = \sum_{m=0}^{\infty} f[m] u[k - m] = \sum_{m=0}^{\infty} f[m].$$

The convolution of two causal signals can be simplified to

$$f_2[k] * f_2[k] = \sum_{m=0}^{k} f_1[m] f_2[k-m]$$
$$= f_1[0] \cdot f_2[k] + f_1[1] \cdot f_2[k-1] + \ldots + f_1[k] \cdot f_2[0]$$

5.2 Introduction to the Z Transform

If we have a continuous, causal signal $f(t)$ and we sample it at an constant time period T, we get a function that we can write as a series of delta functions times numbers representing the amplitude at that point

In Fig. 5.1 we showed a continuous signal sampled at an interval of T seconds, and decided that we could write it as Eq. (5.1). This is still a causal time-domain signal, so we can use the Laplace transform.

$$\mathcal{L}\left\{\sum_{k=0}^{\infty} f_k \delta(t - k \cdot T)\right\} = \sum_{k=0}^{\infty} f[k] e^{-ksT}.$$

By virtue of the time-shifting theorem for Laplace transforms, each time delay T has acquired a term e^{-sT}. T is a constant and $s = \sigma + j\omega$, a complex number. This means that e^{sT} is also a complex number. We will give it a name,

$$z = e^{sT}. \tag{5.8}$$

The one-sided Z transform is,

$$F(z) = \mathcal{Z}\{f[k]\} = \sum_{k=0}^{\infty} f[k] z^{-k}. \tag{5.9}$$

In a way, the Z-transform is the simplest transform because it has the effect of adding z^{-1} for each delay. For instance, for a discrete time function,

$$f[k] = 4\delta[k-2] + 2\delta[k-4] + \delta[k-6],$$

we obtain the Z-domain function,

$$F(z) = 4z^{-2} + 2z^{-4} + z^{-6}.$$

A valuable series formula to remember is the following:

$$1+q+q^2+q^3+\ldots = \frac{1}{1-q} \quad if\ |q|<1. \qquad (5.10)$$

Using Eq. (5.10), we can find the Z transform of the discrete step function.

$$Z\{u[k]\} = 1+z^{-1}+z^{-2}+z^{-3}+z^{-4}\ldots = \frac{1}{1-z^{-1}} = \frac{z}{z-1}$$

Similarly
$$Z\{a^k u[k]\} = 1+az^{-1}+a^2z^{-2}+a^3z^{-3}+a^4z^{-4}\ldots$$
$$= \frac{1}{1-az^{-1}} = \frac{z}{z-a} \qquad (5.11)$$

What is the Z transform of $\delta[k]$? Obviously, this only has one term,

$$Z\{\delta[k]\} = \sum_{k=0}^{\infty} \delta[k]z^{-k} = 1.$$

5.2.1. Properties of Z Transforms

Property 1. Linearity

$$Z\{\alpha f_1[k]+\beta f_2[k]\} = \alpha F_1(z)+\beta F_2(z)$$

The proof comes directly from the definition

$$Z\{\alpha f_1[k]+\beta f_2[k]\} = \sum_{k=0}^{\infty} \{\alpha f_1[k]+\beta f_2[k]\}z^{-k}$$
$$= \sum_{k=0}^{\infty} \alpha f_1[k]z^{-k} + \sum_{k=0}^{\infty} \beta f_2[k]z^{-k}$$
$$= \alpha F_1(z)+\beta F_2(z)$$

Property 2. Right Time Shift

$$Z\{f[k-k_0]u[k-k_0]\} = z^{-k_0}F(z) \quad \text{a } k_0 \text{ positive integer}$$
(5.2.5)

Proof

$$Z\{f[k-k_0]u[k-k_0]\} = \sum_{k=0}^{\infty} z^{-k}f[k-k_0]u[k-k_0]$$

Substitute $i = k - k_0$ or $k = i + k_0$.

$$\sum_{\substack{i+k_0=0 \\ i=-k_0}}^{\infty} z^{-i} z^{-k_0} f[i] u[i] = z^{-k_0} \sum_{i=0}^{\infty} z^{-i} f[i] = z^{-k_0} F(z),$$

because the terms between $-k_0$ and 0 fall out.

Example 5.2. Find the Z transforms of:

$$f_1[k] = (0.5)^k u[k], \qquad f_2[k] = (0.5)^{k-1} u[k-1]..$$

These two functions are illustrated in Fig. (5.6).

Solution

The Z transform of the first is simply

$$F_1(z) = \frac{z}{z - 1/2}.$$

Clearly the second function is merely the first delayed by one so,

$$F_2(z) = z^{-1} F_1(z) = \frac{1}{z - 1/2}.$$

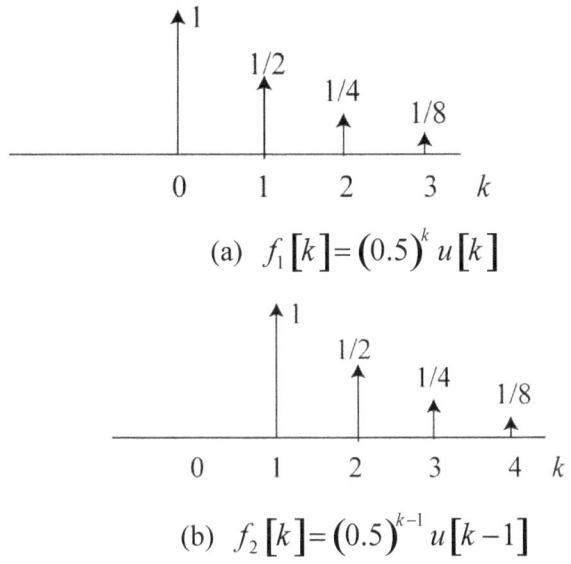

Figure 5.6. Graphs of $f_1[k]$ and $f_2[k]$

Property 2.a Right Time Shift (causal form)

Be careful! There is another version of the right time shift theorem.

$$Z\{f[k-k_0]u[k]\} = z^{-k_0}F(z) + z^{-k_0}\left\{\sum_{i=1}^{k_0} f[-i]z^i\right\}$$

Notice that the difference is that the step function u[k] does not change. This is important, because the Z transform always starts at $k = 0$.

Proof

$$Z\{f[k-k_0]u[k]\} = \sum_{k=0}^{\infty} f[k-k_0]u[k]z^{-k}$$

Change variables $\qquad i = k - k_0$

$$Z\{f[k-k_0]u[k]\} = \sum_{i=-k_0}^{\infty} f[i]u[i+k_0]z^{-(i+k_0)}$$

$$= z^{-k_0} \sum_{i=-k_0}^{\infty} f[i]z^{-i} = z^{-k_0}\left(\sum_{i=0}^{\infty} f[i]z^{-i} + \sum_{i=-k_0}^{-1} f[i]z^{-i}\right)$$

$$= z^{-k_0} F(z) + z^{-k_0} \sum_{i=-k_0}^{-1} f[i]z^{-i}.$$

We will be most interested in shifts of one or two time steps, i.e,

$$Z\{f[k-1]u[k]\} = z^{-1}F(z) + f[-1], \qquad (5.12\ a)$$
$$Z\{f[k-2]u[k]\} = z^{-2}F(z) + z^{-1}f[-1] + f[-2]. \qquad (5.12\ b)$$

Example 5.3. Find the Z transform of

$$f_3[k] = (0.5)^{k-1} u[k]$$

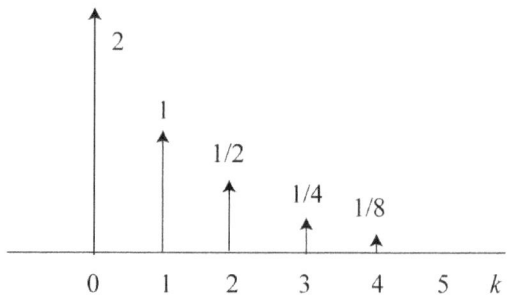

Figure 5.7. Graph of $f_3[k] = (0.5)^{k-1} u[k]$

Solution

Using Eq. (5.12 a),

$$Z\{f[k-1]u[k]\} = Z\{(.5)^{k-1} u[k]\}$$

$$= z^{-1}F(z) + f[-1] = z^{-1}\frac{z}{z-0.5} + 2 = \frac{2z}{z-0.5}$$

From Fig. 5.7 we see that the right shift has brought

$f[-1] = (0.5)^{-1} = 2$ into the picture.

Example 5.4. Solve for the impulse response,

$$y[k] - 0.5y[k-1] = \delta[k].$$

Solution

Start by taking the Z transform of each term. Since the Z transform starts at $n=0$, we are implicitly taking the Z transforms of

$$y[k]u[k] - 0.5y[k-1]u[k] = \delta[k]u[k]$$

$$Y(z) - 0.5\{z^{-1}Y(z) - y[-1]\} = 1$$

Since we are looking for an impulse response, we assume all initial conditions are zero, which leaves

$$Y(z) - 0.5z^{-1}Y(z) = 1,$$

$$Y(z)(1 - 0.5z^{-1}) = 1,$$

$$Y(z) = \frac{1}{1 - 0.5z^{-1}} = \frac{z}{z - 0.5},$$

so

$$h[k] = (0.5)^k u[k].$$

Property 3. Left Time Shift

The general proof is similar to the causal right time shift. Again, we will be most interested in shifts of just one or two.

$$Z\{f[k+1]u[k]\} = zF(z) - zf[0] \quad\quad (5.13\text{ a})$$

$$Z\{f[k+2]u[k]\} = z^2 F(z) - z^2 f[0] - zf[1]. \quad (5.13\text{ b})$$

Example 5.5. Find the Z transform of

$$f_4[k] = (0.5)^{k+1} u[k],$$

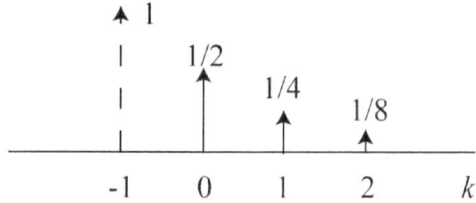

Figure 5.8. Graph of $f_4[k]$.

Solution

Using Eq. (5.12 a),

$$F_2(z) = z\frac{z}{z-0.5} - z$$

$$= \frac{z^2 - z(z-.05)}{z-0.5} = \frac{0.5}{z-0.5}.$$

Notice that the delta function formerly at $k = 0$ has been shifted out of the picture, because the Z transform starts at zero.

Example 5.6 Solve for the impulse response

$$y[k+1] - 0.5y[k] = \delta[k+1].$$

Solution

Notice that this is the same as the previous example, with each function shifted to the left by one time step.

$$zY(z) - zy[0] - 0.5Y(z) = z.$$

Once again, we ignore initial conditions giving,

$$Y(z) = \frac{z}{z-0.5}.$$

Property 4. Convolution

$$Z\{f_1[k] * f_2[k]\} = F_1(z)F_2(z)$$

Proof

$$f_1[k] * f_2[k] = \sum_{m=0}^{\infty} f_1[m] f_2[k-m]$$

$$Z\{f_1[k] * f_2[k]\} = \sum_{k=0}^{\infty} \sum_{m=0}^{\infty} f_1[m] f_2[k-m] z^{-k}$$

$$= \sum_{m=0}^{\infty} f_1[m] \sum_{k=0}^{\infty} f_2[k-m] z^{-k}$$

If f_2 is causal, we can write

$$\sum_{k=0}^{\infty} f_2[k-m] z^{-k} = \sum_{k=0}^{\infty} f_2[k-m] u[k-m] z^{-k},$$

and use the first version of the right shift theorem

$$\sum_{k=0}^{\infty} f_2[k-m] u[k-m] z^{-k} = F_2(z) z^{-m}$$

Then the original equation becomes

$$Z\{f_1[k] * f_2[k]\} = \sum_{m=0}^{\infty} f_1[m] F_2(z) z^{-m}$$

$$= F_2(z) \sum_{m=0}^{\infty} f_1[m] z^{-m}$$

$$= F_1(z) F_2(z).$$

Example 5.7. Convolve the following two signals

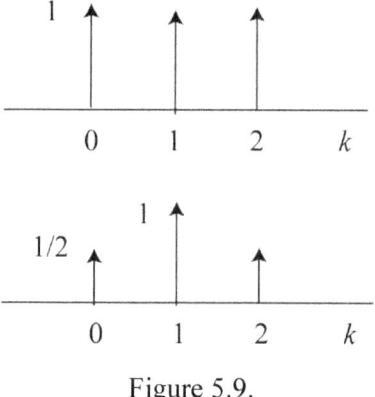

Figure 5.9.

Solution

$$F_1(z) = 1 + z^{-1} + z^2$$

$$F_2(z) = 0.5 + z^{-1} + 0.5z^2$$

$$F_1(z)F_2(z) = (1 + z^{-1} + z^{-2})(0.5 + z^{-1} + 0.5z^{-2})$$
$$= 0.5 + (1 + 0.5)z^{-1} + (0.5 + 0.5 + 1)z^{-2}$$
$$+ (1 + 0.5)z^{-3} + 0.5z^{-4}$$
$$= 0.5 + 1.5z^{-1} + 2z^{-2} + 1.5z^{-3} + 0.5z^{-4}$$

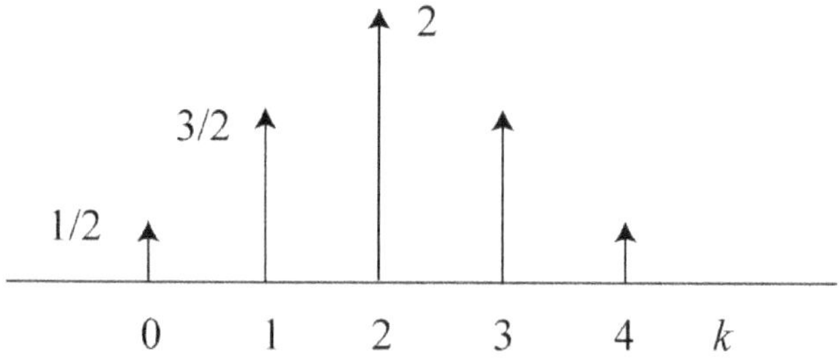

Figure 5.10. Convolution of the two functions in Fig. 5.9.

It is not difficult to see that if we used the same procedure from Example 5.1, we get the same result.

Notice that with the one basic Z transform Eq. (5.11),

$$Z\{a^k u[k]\} = \frac{z}{z-a},$$

we can begin to generate many more transforms. For instance, there is nothing that says *a* cannot be a complex number. So

$$Z\{e^{ik\omega T} u[k]\} = \frac{z}{z - e^{i\omega T}},$$

and

$$Z\{e^{-ik\omega T} u[k]\} = \frac{z}{z - e^{-i\omega T}}.$$

Do not be confused! The ω in this case is some constant value. It is not a variable like it was with the Fourier transforms. Similarly, T is a constant. It is the sampling rate being used by whatever digital system we are dealing with. So, if ω is radians per second and T is in seconds, the quantity ωT is *radians*.

Example 5.8. Find the Z transform of $\cos(k\omega T)$

$$Z\{\cos(k\omega T) u[k]\} = Z\{0.5(e^{ik\omega T} + e^{-ik\omega T}) u[k]\}$$

$$= \frac{1}{2}\left[\frac{z}{z - e^{i\omega T}} + \frac{z}{z - e^{-i\omega T}}\right] = \frac{1}{2}\left[\frac{z(z - e^{-i\omega T}) + z(z - e^{i\omega T})}{z^2 - 2z\cos(\omega T) + 1}\right]$$

$$= \frac{z^2 - z\cos(\omega T)}{z^2 - 2z\cos(\omega T) + 1}.$$

What is the Z transform of $f[k] = a^k \cos(k\omega T) u[k]$?

$$f[k] = a^k \cos(k\omega T) u[k]$$

$$= \frac{1}{2}\left(a^k e^{ik\omega T} + a^k e^{-ik\omega T}\right)$$

$$F(z) = \frac{1}{2}\left(\frac{z}{z+(ae^{i\omega T})} + \frac{z}{z+(ae^{-i\omega T})} \right)$$

$$= \frac{z^2 - za\cos(\omega T)}{z^2 + a2\cos(\omega T) + a^2}.$$

Property 5. Frequency Scaling Property

$$Z\{a^k f[k]\} = F\left(\frac{z}{a}\right) \quad (5.2.8)$$

Proof

$$Z\{a^k f[k]\} = \sum_{k=0}^{\infty} a^k f[k] z^{-k}$$

$$= \sum_{k=0}^{\infty} f[k]\left(\frac{z}{a}\right)^{-k} = F\left(\frac{z}{a}\right).$$

Example 5.9. Find the Z transform of

$$f[k] = (0.5e^{j\omega T})^k u[k]$$

Solution

By frequency scaling

$$F\left(\frac{z}{e^{j\omega T}}\right) = \frac{\frac{z}{e^{j\omega T}}}{\frac{z}{e^{j\omega T}} - 0.5} = \frac{z}{z - 0.5e^{-j\omega T}}.$$

Property 6. Time Multiplication

$$Z\{kf[k]\} = -z\frac{d}{dz}F(z) \quad (5.14)$$

Proof

Start with the definition

$$F(z) = \sum_{n=0}^{\infty} f[k] z^{-k}$$

Take the derivative of both sizes with respect to z

$$\frac{dF(z)}{dz} = \sum_{n=0}^{\infty} -kf[k] z^{-k-1}$$

and multiply both sides by $-z$

$$-z \frac{dF(z)}{dz} = \sum_{n=0}^{\infty} kf[k] z^{-k} = Z\{kf[k]\}$$

Example 5.10. (Problem 5.5 in Gajic)

$$f[k] = 2^k u[k]$$

Find the Z transform of

$$kf[k] + 3^k f[k] + f(k-2) u[k-2] + f[k+1] u[k]$$

<u>Solution</u>

$$F(z) = \frac{z}{z-2}$$

The first term is solved by time multiplication,

$$Z\{kf[k]\} = -z \frac{d}{dz}\left(\frac{z}{z-2}\right) = -z \left(\frac{(z-2)-z(1)}{(z-2)^2}\right) = \frac{2z}{(z-2)^2}.$$

The second term comes from frequency scaling,

$$F\left(\frac{z}{3}\right) = \frac{(z/3)}{(z/3)-2} = \frac{z}{z-6}.$$

The third term from the right shift theorem,

$$Z\{f[k-2] u[k-2]\} = z^{-2} \frac{z}{z-2} = \frac{z^{-1}}{z-2}.$$

The last term comes from the left shift theorem

$$Z\{f[k+1]u[k]\} = z\frac{z}{z-2} - zf[0]$$

$$= z\frac{z}{z-2} - z \cdot 1 = \frac{z^2 - z(z-2)}{z-2} = \frac{2z}{z-2}.$$

5.3. Inverse Z Transforms

The inverse Z transform is defined as follow:

$$f[k] = \frac{1}{2\pi j}\oint_\Gamma F(z)z^{k-1}dz \qquad (5.15)$$

This can be done using contour integration. Needless to say, what we prefer to do is get every term in a form that we can look up in Table 5.1, possibly using the properties of Table 5.2.

Example 5.11. Find the inverse Z transform of

$$F(z) = 1 + 2z^{-2} + 3z^{-4}.$$

Solution

$$f[k] = \delta[k] + 2\delta[k-2] + 3\delta[k-4],$$

or we might say

$$f[0] = 1, \quad f[2] = 2, \quad f[4] = 3.$$

Example 5.12. Solve

$$y[k] - 0.3y[k-1] = u[k], \qquad y[-1] = 0.$$

Solution

First take the Z transforms of both sides

$$Y(z) - 0.3z^{-1}Y(z) = \frac{z}{z-1}$$

$$Y(z) = \frac{z}{(1-0.3z^{-1})(z-1)}$$

$$Y(z) = \frac{z^2}{(z-0.3)(z-1)}$$

In order to get back to the sampled time domain, we have to do a partial fraction expansion. We use a version of partial fraction expansion similar to how we used the Laplace transforms. Start by dividing everything by z. The reason will be apparent shortly.

$$\frac{Y(z)}{z} = \frac{z}{(z-0.3)(z-1)} = \frac{A}{(z-0.3)} + \frac{B}{(z-1)}.$$

Multiplying through by z-0.3 and evaluating at z = 0.3 gives us A,

$$A = \frac{z}{(z-1)}\bigg|_{z=0.3} = \frac{0.3}{-0.7} = -0.429.$$

Multiplying through by z-1 and evaluating at z = 1 gives us B,

$$B = \frac{z}{(z-0.3)}\bigg|_{z=1} = \frac{1}{.7} = 1.429.$$

Now we multiply back through by the z that we saved to give us terms that we can find in the tables:

$$Y(z) = \frac{-0.429z}{(z-0.3)} + \frac{1.429z}{(z-1)}.$$

We can easily go to the sampled time-domain

$$y[k] = \left[1.429 - 0.429(0.3)^k\right]u[k].$$

Check:

$$y[0] = \left[1.429 - 0.429(0.3)^0\right]u[k] = 1$$

$$y[1] = \left[1.429 - 0.429(0.3)^1\right]u[k] = 1.3$$

$$y[2] = \left[1.429 - 0.429(0.3)^2\right]u[k] = 1.39$$

We can compare these values by looking at the original equation:

$$y[k] = 0.3y[k-1] + u[k] \qquad y[-1] = 0$$

At k = 0 $\qquad y[0] = 0.3y[-1] + u[0] = 1$

At k = 1 $\qquad y[1] = 0.3y[0] + u[1] = 1.3$

At k = 2 $\qquad y[2] + 0.3y[1] + u[k] = (0.3)(1.3) + 1 = 1.39$

Example 5.13. Find the inverse Z transform of

$$F(z) = \frac{2z(z - 0.25)}{(z - 0.2)(z - 0.3)}.$$

Solution

We start by dividing out a z so we will have it at the end

$$\frac{F(z)}{z} = \frac{2(z - 0.25)}{(z - 0.2)(z - 0.3)} = \frac{A}{z - 0.2} + \frac{B}{z - 0.3}, \qquad (5.16)$$

and then solve for A and B:

$$A = \left.\frac{2(z - 0.25)}{(z - 0.3)}\right|_{Z=0.2} = \frac{2(0.2 - 0.25)}{(0.2 - 0.3)} = \frac{2(-.05)}{-.1} = 1$$

$$B = \left.\frac{2(z - 0.25)}{(z - 0.2)}\right|_{Z=0.3} = \frac{2(0.3 - 0.25)}{(0.3 - 0.2)} = \frac{2(.05)}{.1} = 1$$

$$F(z) = \frac{z}{z - 0.2} + \frac{z}{z - 0.3}$$

Check the partial fraction expansion:

$$F(z) = \frac{z}{z - 0.2} + \frac{z}{z - 0.3} = \frac{z(z - 0.3) + z(z - 0.2)}{(z - 0.3)(z - 0.2)} = \frac{2z^2 + 0.5z}{(z - 0.3)(z - 0.2)}$$

The inverse Z is:

$$f[k] = \left[(0.2)^k + (0.3)^k\right] u[k]$$

Example 5.14. Find the inverse Z transform of

$$F(z) = \frac{1}{2z-1}.$$

Solution

First we will rewrite F(z) as

$$F(z) = \frac{0.5}{z-0.5}.$$

Remember, we need to divide out a z term to find the inverse. One solution is to just multiply both sides by $1/z$, and then do a partial fraction expansion:

$$\frac{F(z)}{z} = \frac{0.5}{z(z-0.5)} = \frac{A}{z} + \frac{B}{z-0.5}$$

$$A = \left.\frac{0.5}{z-0.5}\right|_{z=0} = -1, \quad B = \left.\frac{0.5}{z}\right|_{z=0.5} = 1$$

$$F(z) = \frac{-z}{z} + \frac{z}{z-0.5}$$

$$f[k] = -\delta[k] + (0.5)^k u[k]$$

$$f[0] = -1 + (0.5)^0 = 0$$

$$f[1] = 0.5$$

$$f[2] = 0.25$$

Notice that this is $(0.5)^k u[k]$ with the first term missing. Could we have done this another way? Suppose we say

$$F(z) = z^{-1} \frac{0.5z}{z-.5}.$$

We know

$$Z^{-1}\left\{\frac{0.5z}{z-0.5}\right\} = 0.5(0.5)^k u[k].$$

We can account for the z^{-1} that we left out, by shifting to the right in the sampled time domain,

$$f[k] = 0.5(0.5)^{k-1} u[k-1]$$
$$= (0.5)^k u[k-1].$$

Notice that

$$f[0] = 0$$
$$f[1] = 0.5$$
$$f[2] = 0.25,$$

which is in agreement with the result we got by multiplying by 1/z and doing a partial fraction expansion.

Example 5.15. Find the inverse Z transform of

$$F(z) = \frac{1}{(z-1)(z-0.5)}.$$

Solution

Once again, we do not have a z term in the numerator to divide out and use later. We could multiply both sides by 1/z, but that adds an additional term and makes the problem more complicated. Instead, we write the following:

$$F(z) = z^{-1}\left[\frac{z}{(z-1)(z-0.5)}\right] = \frac{A}{z-1} + \frac{B}{z-0.5},$$

and we begin by taking the partial fraction expansion of the term in the brackets:

$$F_0(z) = \left[\frac{z}{(z-1)(z-0.5)}\right] = \frac{A}{z-1} + \frac{B}{z-0.5}.$$

$$A = \frac{1}{(z-0.5)}\bigg|_{z=1} = 2, \quad B = \frac{1}{(z-1)}\bigg|_{z=0.5} = -2.$$

Now we get $F(z)$ from $F_0(z)$

$$F(z) = z^{-1} F_0(z) = z^{-1}\left(\frac{2z}{z-1}\right) - z^{-1}\left(\frac{2z}{z-0.5}\right).$$

The terms in the parentheses can be found in the table. Then we apply the right shift theorem:

$$f[k] = \left(2 - 2(0.5)^{k-1}\right) u[k-1].$$

Example 5.16. Find the inverse Z transform of

$$F(z) = \frac{z^2 + 2z - 1}{z^2 + 3z + 2}.$$

Solution

Notice that no term in the Table 5.1 looks like $F(z)$. We start by dividing.

$$F(z) = \frac{z^2 + 2z - 1}{z^2 + 3z + 2} = 1 + \frac{-z - 3}{z^2 + 3z + 2}.$$

The inverse of one is just the delta function. We can take the partial fraction of the second term after we realize that we are missing a z in the numerator that we can divide out. Therefore, we will remember that we owe a z^{-1}, and right shift our time domain answer.

$$F_0(z) = \frac{-z - 3}{z^2 + 3z + 2} = \frac{A}{z+1} + \frac{B}{z+2}$$

$$A = \frac{-z-3}{z+2}\bigg|_{z=-1} = \frac{-2}{1} = -2, \quad B = \frac{-z-3}{z+1}\bigg|_{z=-2} = \frac{-1}{-1} = 1.$$

$$F(z) = 1 - \frac{2}{z+1} + \frac{1}{z+2}.$$

$$f[k] = \delta[k] - 2(-1)^{k-1} u[k-1] + (-2)^{k-1} u[k-1]$$

213

Example 5.17. Solve the following difference equation. Assume all initial conditions are zero.

$$y[k+2] - 0.3y[k+1] + 0.02y[k] = (0.01)(0.3)^k u[k]$$

Solution

Take the Z transform

$$z^2 Y(z) - 0.3zY(z) + 0.02Y(z) = 0.01 \frac{z}{z - 0.3}$$

$$Y(z)(z^2 - 0.3z + 0.02) = 0.01 \frac{z}{z - 0.3}$$

$$Y(z) = \frac{0.01z}{(z - 0.1)(z - 0.2)(z - 0.3)}$$

$$\frac{Y(z)}{z} = \frac{A}{(z - 0.1)} + \frac{B}{(z - 0.2)} + \frac{C}{(z - 0.3)}$$

$$A = \frac{0.01}{(z - 0.2)(z - 0.3)}\bigg|_{z=0.1} = \frac{0.01}{(-0.1)(-0.2)} = 0.5$$

$$B = \frac{0.01}{(z - 0.1)(z - 0.3)}\bigg|_{z=0.2} = \frac{0.01}{(0.1)(-0.1)} = -1$$

$$C = \frac{0.01}{(z - 0.1)(z - 0.2)}\bigg|_{z=0.3} = \frac{0.01}{(0.2)(0.1)} = 0.5$$

$$y[k] = \left(0.5(0.1)^k - (0.2)^k + 0.5(0.3)^k\right) u[k]$$

We can check the answer. Looking at the original equation, the first nonzero term occurs for

k = 0 $y[2] - 0.3y[1] + 0.02y[0] = (0.01)$, so $y[2] = 0.001$

k = 1 $y[3] - 0.3(0.01) = (0.01)(.3)$ so $y[3] = 0.006$

Now look at the analytic solution:
$$y[0] = (0.5 - 1 + 0.5) = 0$$
$$y[1] = (0.5(0.1) - (0.2) + 0.5(0.3)) = (0.05 - 2 + 0.15) = 0$$
$$y[2] = (0.5(0.01) - (0.04) + 0.5(0.09))$$
$$= 0.005 - 0.04 + .045 = 0.001$$
$$y[3] = (0.5(0.1)^3 - (0.2)^3 + 0.5(0.3)^3)$$
$$= .0005 - .008 + .0135 = 0.006$$

5.3.1. Partial Fraction Expansion of Multiple Poles

Find the inverse Z transform of
$$F(z) = \frac{z(z+1)}{(z-1)(z-.25)^2}.$$

The first step is to save a z:
$$\frac{F(z)}{z} = \frac{(z+1)}{(z-1)(z-.25)^2} = \frac{A}{z-1} + \frac{B}{(z-.25)^2} + \frac{C}{(z-.25)} \qquad (5.17)$$

To find A, multiply through by z-1 and evaluate at z = 1,
$$A = \frac{(z+1)}{(z-.25)^2}\bigg|_{z=1} = \frac{2}{(.75)^2} = 3.56.$$

To find B, multiply through by $(z-.25)^2$
$$\frac{(z+1)}{(z-1)} = \frac{A}{z-1}(z-.25)^2 + B + C(z-.25), \qquad (5.18)$$

and then evaluate at z = .25
$$B = \frac{(z+1)}{(z-1)}\bigg|_{z=.25} = \frac{1.25}{-.75} = -\frac{5\ 4}{4\ 3} = -\frac{5}{3} = -1.67.$$

To get C, we first take the derivative of Eq. (5.18) with respect to z

$$\frac{d}{dz}\left[\frac{(z+1)}{(z-1)}\right] = \frac{d}{dz}\left[\frac{A}{z-1}(z-.25)^2\right] + C,$$

and then evaluate at z = .25, which will eliminate the term with the A,

$$C = \frac{d}{dz}\left[\frac{(z+1)}{(z-1)}\right]_{z=.25} = \frac{1(z-1)-1(z+1)}{(z-1)^2}$$

$$= \frac{-2}{(z-1)^2}\bigg|_{z=.25} = \frac{-2}{(-.75)^2} = -3.56$$

$$F(z) = \frac{(z+1)}{(z-1)(z-.25)^2} = \frac{3.56z}{z-1} + \frac{-1.67z}{(z-.25)^2} + \frac{-3.56z}{(z-.25)}$$

The second term we rewrite to match a formula in Z transform table:

$$-1.67\frac{z}{(z-.25)^2} = \frac{-1.67}{.25}\frac{0.25z}{(z-.25)^2} = -6.68\frac{0.25z}{(z-.25)^2}.$$

Therefore,

$$f[k] = \left[3.56 - 6.68 \cdot k(.25)^k - 3.56(.25)^k\right]u[k]$$

Example 5.18. Take the partial fraction expansion of

$$F(z) = \frac{z(z+1)}{(z-1)(z-.25)^3}.$$

Solution
First, save a z:

$$\frac{F(z)}{z} = \frac{(z+1)}{(z-1)(z-.25)^3} = \frac{A}{z-1} + \frac{B}{(z-.25)^3} + \frac{C}{(z-.25)^2} + \frac{D}{(z-.25)}$$

We get A in the usual way:

$$A = \left.\frac{(z+1)}{(z-.25)^3}\right|_{z=1} = 4.74$$

We will solve for B, C, and D by multiplying through by $(z-.25)^3$

$$\frac{(z+1)}{(z-1)} = \frac{A}{z-1}(z-.25)^3$$
$$+ B + C(z-.25)^1 + D(z-.25)^2 \qquad (5.19)$$

We get B by evaluating at z=0.25

$$B = \left.\frac{(z+1)}{(z-1)}\right|_{z=1/4} = \frac{5/4}{-3/4} = -0.6.$$

To get C, we take the first derivative of Eq. (5.19),

$$\frac{d}{dz}\frac{(z+1)}{(z-1)} = \frac{A}{z-1} 3(z-.25)^2 + C + D2(z-.25),$$

and then evaluate at z = ¼:

$$C = \left.\frac{d}{dz}\frac{(z+1)}{(z-1)}\right|_{z=1/4} = \left.\left[\frac{1(z-1)-1(z+1)}{(z-1)^2}\right]\right|_{z=1/4}$$

$$= \left.\left[\frac{-2}{(z-1)^2}\right]\right|_{z=1/4} = -3.55$$

To get D, we have to take another derivative of Eq. (5.19),

$$\frac{d^2}{dz^2}\left[\frac{(z+1)}{(z-1)}\right] = \frac{A}{z-1} 3 \cdot 2(z-.25)^1 + D2,$$

and evaluate at z = ¼:

$$D = \frac{1}{2}\frac{d^2}{dz^2}\left[\frac{(z+1)}{(z-1)}\right]_{z=1/4} = \frac{1}{2}\left[\frac{d}{dz}\frac{-2}{(z-1)^2}\right]_{z=1/4}$$

$$= \frac{1}{2}\left[\frac{4}{(z-1)^3}\right]_{z=1/4} = 4.74$$

Now we have all the constants in Eq. (5.19):

$$F(z) = \frac{4.74z}{z-1} + \frac{-0.6z}{(z-.25)^3} + \frac{-3.55z}{(z-.25)^2} + \frac{4.74z}{(z-.25)}.$$

Example 5.19. Find the inverse Z transform of,

$$F(z) = \frac{1}{(z-0.5)(z-1)^2}.$$

Solution

Note that we are missing a z in the numerator to save, but we will proceed with the expansion:

$$F(z) = \frac{1}{(z-0.5)(z-1)^2} = \frac{A}{(z-0.5)} + \frac{B}{(z-1)^2} + \frac{C}{z-1}.$$

$$A = \frac{1}{(z-1)^2}\bigg|_{z=0.5} = 4, \quad B = \frac{1}{(z-0.5)}\bigg|_{z=1} = 2,$$

$$C = \frac{d}{dz}\frac{1}{(z-0.5)}\bigg|_{z=2} = \frac{-1}{(z-0.5)^2}\bigg|_{z=2} = -4.$$

Now we write

$$F(z) = z^{-1}\left(\frac{4z}{(z-0.5)}\right) + z^{-1}\left(\frac{2z}{(z-1)^2}\right) - z^{-1}\left(\frac{4z}{z-1}\right).$$

When we go back to the sample time domain, we get

$$f[k] = \left(4(0.5)^{k-1} + 2(k-1) - 4\right)u(k-1).$$

5.4. Inverse Z Transforms of Complex Poles

Just as with the Laplace transforms, we will have occasion to look for the inverse Z transforms of complex conjugate pairs. Once again, we start with the cosine method because it follows intuitively. But it will prove more efficient to use the sine method.

5.4.1 The Cosine Method

Suppose we want to find the inverse Z transform of

$$F(z) = \frac{2z^2}{z^2 - \sqrt{3}z + 1}.$$

We start by rewriting

$$\frac{F(z)}{z} = \frac{2z}{z^2 - \sqrt{3}z + 1} = \frac{c}{z-p} + \frac{c^*}{z-p^*}. \qquad (5.20)$$

We find the ps by solving

$$p = \frac{\sqrt{3} + \sqrt{3-4}}{2} = \frac{\sqrt{3}}{2} + j\frac{1}{2} = \frac{2\angle 30°}{2} = 1\angle 30°$$

$$p^* = \frac{\sqrt{3}}{2} - j\frac{1}{2} = 1\angle -30°$$

Remember: We follow the convention that p is always calculated using the plus sign on the imaginary term.

We find c by

$$c = \left.\frac{2z}{z-p^*}\right|_{z=p} = \frac{2\left(\frac{\sqrt{3}}{2} + j\frac{1}{2}\right)}{\left(\frac{\sqrt{3}}{2} + j\frac{1}{2}\right) - \left(\frac{\sqrt{3}}{2} - j\frac{1}{2}\right)}.$$

$$= \frac{\sqrt{3}+j}{j} = \frac{\sqrt{4}\angle 30°}{1\angle 90°} = 2\angle -60°.$$

Not surprisingly, when we solve for c^* we get the complex conjugate:

$$c^* = \left.\frac{2z}{z-p}\right|_{z=p^*} = \frac{2\left(\frac{\sqrt{3}}{2} - j\frac{1}{2}\right)}{\left(\frac{\sqrt{3}}{2} - j\frac{1}{2}\right) - \left(\frac{\sqrt{3}}{2} + j\frac{1}{2}\right)}$$

$$= \frac{\sqrt{3} - j}{-j} = \frac{\sqrt{4}\angle -30^o}{1\angle -90^o} = 2\angle 60^o.$$

Now go back and put the two together

$$\frac{F(z)}{z} = \frac{c}{z-p} + \frac{c^*}{z-p^*} = \frac{2e^{-j60^o}}{z - e^{j30^o}} + \frac{2e^{j60^o}}{z - e^{-j30^o}},$$

or

$$F(z) = \frac{z2e^{-j60^o}}{z - e^{j30^o}} + \frac{z2e^{j60^o}}{z - e^{-j30^o}}.$$

The inverse Z transform gives

$$f[k] = 2e^{-j60^o}\left(e^{j30^o}\right)^k + 2e^{j60^o}\left(e^{-j30^o}\right)^k u[k]$$

$$= 2 \cdot (1)^k \left[e^{j(k30^o - 60^o)} + e^{-j(k30^o - 60^o)}\right] u[k]$$

$$= 4 \cdot (1)^k \cos(30^o k - \angle 60^o) u[k]$$

The generalization of this formula would be

$$f[k] = 2|c||p|^k \cos(k\angle p + \angle c) \qquad (5.21)$$

5.4.2 The Sine Method

In looking at inverse Laplace transforms with conjugate coefficients, we found a shortcut that we referred to as the sine method. Let us see if we can find a similar shortcut here. First, look back on the calculation of c

$$c = \left.\frac{2z}{z-p^*}\right|_{z=p} = \frac{2p}{p-p^*} = \frac{\sqrt{3}+j}{j} = \frac{\sqrt{4}\angle 30^o}{1\angle 90^o}.$$

The numerator is a pure imaginary number, because when we take the difference between a number and its complex conjugate, we will always get two times the imaginary part of the original number

$$p - p^* = (\alpha + j\beta) - (\alpha - j\beta) = 2j\beta.$$

Knowing this, we look for a shortcut. First, write c as

$$c = \left.\frac{2z}{z-p^*}\right|_{z=p} = \frac{d}{2j\beta} = \frac{2p}{2j\beta} = \frac{\sqrt{4}\angle 30°}{2j\beta}.$$

Since we know the $j2\beta$ will always be in the denominator, we could have just solved for

$$d = 2z|_{z=p} = \sqrt{3} + j = \sqrt{4}\angle 30°.$$

Similarly

$$c^* = \left.\frac{2z}{z-p}\right|_{z=p^*} = \frac{2p^*}{-2j\beta} = \frac{\sqrt{3}+j}{-2j\beta} = \frac{\sqrt{4}\angle 30°}{-2j\beta},$$

or we could have just calculated

$$d = 2z|_{z=p} = \sqrt{3} + j = \sqrt{4}\angle 30°,$$

where we had calculated $\beta = 1/2$. When we take the inverse of the two terms we get

$$f[k] = \frac{1}{2j\beta}\left[2e^{j\angle 30°}\left(2e^{j\angle p}\right)^k - 2e^{j\angle -30°}\left(2e^{-j\angle p}\right)^k\right]$$

$$= \frac{1}{\beta}\left[2\left(2^k\right)\sin\left(k\angle p + 30°\right)\right]$$

$$= 4\left(2^k\right)\sin\left(k\angle p + 30°\right).$$

We can compare this with our previous result by

$$f[k] = 4\left(2^k\right)\sin\left(30°k + 90° - 90° + 30°\right)$$

$$= 4\left(2^k\right)\cos\left(30°k° - 60°\right)$$

The generalization of this method is

$$f[k] = \frac{1}{\beta}|d||p|^k \sin\left(k\angle p + \angle d\right) \qquad (5.22)$$

The advantage of this sine method is that the calculation of d is slightly simpler than the calculation of c.

5.4.3. Summary

To find the inverse of the complex conjugate component of a function like

$$\frac{F(z)}{z} = \frac{F_1(z)}{z^2 + bz + c} = \frac{F_1(z)}{(z-p)(z-p^*)},$$

use the following sine method:

1. Determine the pole with the positive imaginary part

$$p = \frac{-b + \sqrt{b^2 - 4c}}{2} = \alpha + j\beta = |p|\angle p, \qquad (5.23\ \text{a})$$

2. Calculate

$$d = F_1(z)\big|_{z=p} = |d|\angle d, \qquad (5.23\ \text{b})$$

3. That part of the time domain function due to the complex conjugate pair is

$$f[k] = \frac{1}{\beta}|d||p|^k \sin(k\angle p + \angle d). \qquad (5.23\ \text{c})$$

There is a very simple method to practice using these methods. Look at the decaying sine and cosine functions in the table of Z transforms, put some numbers in them, and make sure you can get back to the original time domain functions.

Example 5.20. Let us start with a term from the table and see if the method really works.

$$Z\left[a^k \sin(\omega T \cdot k)\right] = \frac{za\sin(\omega T)}{z^2 - 2a\cos(\omega T)z + a^2}.$$

Solution

Remember, ωT is an *angle!* Take $a = 0.8$, $\omega T = 30°$. That gives us

$$F(z) = \frac{za\sin(\omega T)}{z^2 - 2a\cos(\omega T)a - a^2} = \frac{0.8(0.5)z}{z^2 - (0.8)(0.866) + (0.8)^2}.$$

So we will solve

$$\frac{F(z)}{z} = \frac{0.4}{z^2 - 1.385z + (0.8)^2},$$

using the sine method:

1. Find the root with the positive imaginary part

$$p = \frac{1.385 + \sqrt{(1.385)^2 - 4(0.64)}}{2}$$

$$= \frac{1.385 + \sqrt{1.92 - 2.56}}{2} = 0.692 + j0.4 = 0.8\angle 30°$$

2. $\quad d = F_1(z)\big|_{z=p} = 0.4\angle 0°$

3. The time-domain function is

$$f_s[k] = \frac{|d|}{\beta}|p|^k \sin[\angle p \cdot k + \angle d] \cdot u[k]$$

$$= \frac{0.4}{0.4}(0.8)^k \sin[30°k] \cdot u[k]$$

Example 5.21. Find the inverse Z transform of

$$F(z) = \frac{z^2 - z}{(z - 0.5)(z^2 + 2z + 5)}.$$

<u>Solution</u>

$$\frac{F(z)}{z} = \frac{z - 1}{(z - 0.5)(z^2 + 2z + 5)} = \frac{A}{(z - .5)} + \frac{M}{(z^2 + 2z + 5)}.$$

We find A in the usual manner:

$$A = \left.\frac{z-1}{(z^2+2z+5)}\right|_{z=1/2} = \frac{-1/2}{1/4+1+5} = \frac{-1}{2}\frac{4}{25} = -\frac{2}{25} = -0.08.$$

We find the second term using the sine method. (Note: the M in the numerator of the second term has no significant meaning. It is just a "place holder.")

$$p = \frac{-2+\sqrt{2^2-4\cdot 5}}{2} = -1+\sqrt{1-5} = -1+j2 = \sqrt{5}\angle 117°.$$

Note that $\beta = 2$.

$$d = \left.\frac{z-1}{(z-0.5)}\right|_{p=-1+j2} = \frac{-2+j2}{-1.5+j2} = \frac{2\sqrt{2}\angle 135°}{2.5\angle 127°} = 1.13\angle 8°$$

Sine term

$$f[k] = \left\{-0.08(0.5)^k + \frac{1.13}{2}\left(\sqrt{5}\right)^k \sin\left[117°k + 8°\right]\right\}u[k].$$

Example 5.22. Find the inverse Z transform of,

$$F_3(z) = \frac{1}{z^2+2z+2}.$$

Solution

Looking back to Eq. (5.23) and Eq. (5.23), we see that implicit to the sine method is a z term in the numerator. Similar to the method we used in Section 5.3, we multiply the Z-domain term by $z^{-1}z$ to provide us with the needed z. Then the z^{-1} results in an application of the right shift theorem at the end. So we start by writing:

$$\frac{F_3(z)}{z} = z^{-1}\left\{\frac{1}{z^2+2z+2}\right\}.$$

Now we use the sine method, Eq. (5.4.3):

i.
$$p = \frac{-2+\sqrt{2^2-4\cdot 2}}{2} = -1+j = \sqrt{2}\angle 135°$$

224

ii. $\quad d = 1$

iii. $\quad f_3'[k] = \frac{1}{1}\left(\sqrt{2}\right)^k \sin[135k]u[k].$

Now we apply the right shift theorem, which results in a delay of one:

$$f_3[k] = \left(\sqrt{2}\right)^{k-1} \sin[135(k-1)]u[k-1].$$

Example 5.23. (Problem 5.12b from Gajic)

Find the inverse Z transform of

$$F_2(z) = \frac{z^2}{(z^2+1/4)(z^2+1/9)}.$$

Solution

We will start by assuming z^2 is the parameter of interest. Begin by writing

$$\frac{F_2(z)}{z^2} = \frac{1}{(z^2+1/4)(z^2+1/9)} = \frac{A}{(z^2+1/4)} + \frac{B}{(z^2+1/9)}$$

$$A = \frac{1}{(z^2+1/9)}\bigg|_{z^2=-1/4} = \frac{1}{(-1/4+1/9)} = \frac{1}{(-9+4)/36} = \frac{36}{-5}$$

$$B = \frac{1}{(z^2+1/4)}\bigg|_{z^2=-1/9} = \frac{36}{5}.$$

Check:

$$\frac{36}{-5}\left(z^2+\frac{1}{9}\right) + \frac{36}{5}\left(z^2+\frac{1}{4}\right) = -\frac{4}{5}+\frac{9}{5} = 1$$

Look at the transform pair

$$a^k \cos(\omega T k)u[k] \qquad \frac{z^2 - az\cos(\omega T)}{z^2 - 2az\cos(\omega T) + a^2}$$

If $\omega T = 90°$, then the transform pair becomes

$$a^k \cos(90°k)u[k] \qquad \frac{z^2}{z^2 + a^2}$$

so with $a = \frac{1}{2}$

$$Z^{-1}\left[\frac{z^2}{(z^2 + 1/4)}\right] = \left(\frac{1}{2}\right)^k \cos(90°k)u[k].$$

And if $a = 1/3$

$$Z^{-1}\left[\frac{z^2}{(z^2 + 1/9)}\right] = \left(\frac{1}{9}\right)^k \cos(90°k)u[k]$$

so

$$f[k] = \frac{36}{5}\left(\frac{1}{3}\right)^k \cos(90°k)u[k] - \frac{36}{5}\left(\frac{1}{2}\right)^k \cos(90°k)u[k]$$

Example 5.24 (Problem 5.19 from Gajic) Derive the formula for the Z-transform of a periodic discrete-time signal defined by
$$f[k] = f[k+N]$$
Solution

$$F(z) = \sum_{k=0}^{\infty} f[k]z^{-k} = \sum_{n=0}^{\infty}\sum_{k=0}^{N-1} f[k]z^{-k}z^{-nN}$$

$$= F_1(z)\sum_{n=0}^{\infty}(z^N)^{-n} = F_1(z)\frac{z^N}{z^N - 1}$$

Example 5.25. Find the inverse Z transform of
$$F_3(z) = \frac{z^2 + 2z + 2}{z^2 - 2z + 4}$$
Solution

If we look in the table of Z transforms, there is no term that is second order in z in the numerator and denominator. Therefore, start by dividing

the numerator by the denominator.
$$F_3(z) = \frac{z^2 + 2z + 2}{z^2 - 2z + 4} = 1 + \frac{4z - 2}{z^2 - 2z + 4}.$$
We apply the sine method to the second term, but notice that we will also have to apply the right shift at the end:
$$p = \frac{2 + \sqrt{4 - 4(4)}}{2} = 1 + j\sqrt{3} = 2\angle 60°$$
$$d = 4z - 2\big|_{z = 1 + j\sqrt{3}} = 2 + j4\sqrt{3} = 7.2\angle 74°$$

$$f_3[k] = \delta[k] + \frac{7.2}{\sqrt{3}} (2)^{k-1} \sin\left[60°(k-1) + 74°\right] u[k-1]$$

5.5. Solving Difference Equations Using Z Transforms

First, review the theorems that are used in solving difference equations:
$$Z\{f[k-1]u[k]\} = z^{-1}F(z) + f[-1]$$
$$Z\{f[k+1]u[k]\} = zF(z) - zf[0]$$

Note the following two important things:

1). The $u[k]$ is implicit to the causality of the Z transform. So if you have $y[k-1]$, assume that it is $y[k-1]u[k]$ and not $y[k-1]u[k-1]$.

2). Initial conditions play no role in taking the Z transform of the forcing function in a difference equation!

Example 5.26. (Time-Forwarded Method)
$$y[k+1] - 0.3y[k] = f[k]$$
$$y[0] = 0.3 \quad f[k] = u[k]$$

Solution

$$zY(z) - zy[0] - 0.3Y(z) = \frac{z}{z - 1}$$

$$Y(z)(z-0.3) = 0.3z + \frac{z}{z-1}$$

$$Y(z) = \frac{z}{z-.3}0.3 + \frac{z}{(z-0.3)(z-1)}$$

We could combine them, but instead we will just solve

$$\frac{Y_0(z)}{z} = \frac{1}{(z-0.3)(z-1)} = \frac{A}{z-0.3} + \frac{B}{z-1}$$

$$A = \frac{1}{z-1}\bigg|_{z=0.3} = \frac{1}{-.7} = -1.43$$

$$B = \frac{1}{z-0.3}\bigg|_{z=1} = \frac{1}{.7} = 1.43$$

$$y[n] = \{0.3(0.3)^k - 1.43(0.3)^k + 1.43\}u[k]$$
$$= \{-1.13(0.3)^k + 1.43\}u[k]$$

Note

$$y[0] = \{-1.13(0.3)^0 + 1.43\}u[k] = 0.3,$$

in keeping with the initial condition that we had.

Example 5.27. (Time-Delayed Method)

$$y[k] - 0.3y[k-1] = f[k-1]$$
$$y[0] = 0.3 \quad f[k] = u[k]$$

As soon as we take the Z transform, we realize we have the wrong initial condition:

$$Y(z) - 0.3\{z^{-1}Y(z) + y[-1]\} = \frac{1}{z-1}$$

Solution

Using the initial condition we have, along with the original

equation, we can find the initial condition we need. For instance, at k = 0
$$y[0]-0.3y[-1]=f[-1],$$
$$0.3-0.3y[-1]=0.$$
This gives the y[-1] = 1 that we needed.
$$Y(z)-0.3\{z^{-1}Y(z)+y[-1]\}=\frac{1}{z-1}$$
$$Y(z)(1-0.3z^{-1})=.3+\frac{1}{z-1}$$
$$Y(z)=\frac{.3}{1-0.3z^{-1}}+\frac{1}{(1-0.3z^{-1})(z-1)}$$
$$=\frac{.3z}{z-0.3}+\frac{z}{(z-0.3)(z-1)}$$
This is obviously the same as the previous example,
$$y[n]=\{-1.13(0.3)^k+1.43\}u[k].$$

Example 5.28. Solve
$$y[k+1]+.3y[k]=f[k]$$
$$f[k]=\delta[k] \quad y[1]=0.7$$

Solution
$$zY(z)-zy[0]+.3Y(z)=1$$
We need to find $y[0]$. The original equation at k = 0 is
$$y[1]+.3y[0]=1$$
$$0.7+.3y[0]=1 \implies y[0]=1$$
Now we can solve
$$zY(z)-z+.3Y(z)=1,$$

$$Y(z) = \frac{z+1}{z+.3} = 1 + \frac{.7}{z+.3},$$

$$y[k] = \delta[k] + 0.7(-.3)^{k-1} u[k-1].$$

We can check the answers by seeing if the difference equation leads to the same result. We know that $y[0] = 1, y[1] = 0.7$. After that,

$$y[2] = -.3y[1] = -.21$$

$$y[3] = -.3y[2] = .063$$

The solution gives $y[0] = 1$, and after that

$$y[1] = 0.7(-.3)^{k-1} = 0.7$$

$$y[2] = 0.7(-.3)^{1} = -0.21$$

$$y[3] = 0.7(-.3)^{2} = 0.063$$

Example 5.29. The following three examples are related so we will refer to them as A, B, and C.

Example A Solve

$$y[k+2] - .8y[k+1] - 0.2y[k] = f[k]$$

$$y[0] = 1, y[-1] = 0$$

For the zero input response (i.e., no forcing function).

Solution

$$z^2 Y(z) - z^2 y[0] - zy[1] - .8\{zY(z) - zy[0]\} - .2Y(z) = 0$$

We do not have y[1]. Look at the original equation for k=-1

$$y[1] - .8y[0] - 0.2y[-1] = 0,$$

$$y[1] - .8(1) - 0.2(0) = 0.$$

So clearly

$$y[1] = 0.8,$$

and the original Z domain equation becomes

$$z^2 Y(z) - z^2(1) - z(0.8) - .8\{zY(z) - z(1)\} - .2Y(z) = 0$$

$$Y(z)\{z^2 - .8z - 0.2\} = z^2$$

$$z = \frac{.8 \pm \sqrt{.64 - 4(-.2)}}{2} = \frac{.8 \pm 1.2}{2} = .4 \pm .6 = -.2, 1$$

$$Y(z) = \frac{z^2}{(z+.2)(z-1)}$$

$$\frac{Y(z)}{z} = \frac{z}{(z+.2)(z-1)} = \frac{A}{z+.2} + \frac{B}{z-1}$$

$$A = \left.\frac{z}{(z-1)}\right|_{z=-.2} = \frac{-0.2}{-1.2} = 0.167$$

$$B = \left.\frac{z}{(z+0.2)}\right|_{z=1} = \frac{1}{1.2} = 0.833$$

$$y_{zi}[k] = (0.833 + 0.167(-.2)^k)u[k]$$

Example B. Solve

$$y[k+2] - .8y[k+1] - 0.2y[k] = f[k]$$

$$f[k] = \delta[k] \qquad y[0] = 1, y[1] = 0.8$$

This is the same as Example A but it also has a forcing function.

Solution

$$z^2 Y(z) - z^2 y[0] - zy[1] - .8\{zY(z) - zy[0]\} - .2Y(z) = 1$$

$$Y(z)\{z^2 - .8z - 0.2\} = z^2 + 1$$

$$Y(z) = \frac{z^2 + 1}{(z+.2)(z-1)}.$$

We need a z to divide through and save so we might say

$$\frac{Y(z)}{z} = \frac{z^2 + 1}{z(z+.2)(z-1)} = \frac{A}{z} + \frac{B}{z+.2} + \frac{C}{z-1}.$$

But stop and look at the alternative formulation:

$$Y(z) = \frac{z^2 + 1}{(z+.2)(z-1)} = \frac{z^2}{(z+.2)(z-1)} + \frac{1}{(z+.2)(z-1)}.$$

The first part we already know from the previous example:

$$y_{zi}[k] = \left(0.833 + 0.167(-.2)^k\right) u[k]$$

The second part can be found from the following:

$$\frac{Y_{zs}(z)}{z} = \frac{C}{(z+.2)} + \frac{D}{(z-1)}$$

$$C = \frac{1}{z-1}\bigg|_{z=-.2} = 5 \quad D = \frac{1}{1.2} = 0.833$$

$$y_{zs}[k] = \left(0.833 + 5(-.2)^{(k-1)}\right) u(k-1)$$
$$= \left(0.833 + 5(-.2)^k\right) u(k) - (0.833 + 5)\delta(k)$$

$$y_{total}[k] = y_{zi}[k] + y_{zs}[k]$$
$$= \left(0.833 + 0.167(-.2)^k\right) u[k]$$
$$+ \left(0.833 + 5(-.2)^k\right) u(k) - (0.833 + 5)\delta(k)$$
$$= \left(1.67 + 5.167(-.2)^k\right) u[k] - (0.833 + 5)\delta(k)$$

Example C. Solve

$$y[k+2] - .8y[k+1] - 0.2y[k] = f[k]$$
$$f[k] = u[k] \quad y[0] = 1, y[1] = 0.8$$

Solution

$$z^2 Y(z) - z^2 y[0] - zy[1] - .8\{zY(z) - zy[0]\} - .2Y(z) = \frac{z}{z-1}$$

$$Y(z)\{z^2 - .8z - 0.2\} = z^2 + \frac{z}{z-1}$$

$$Y(z) = \frac{z^2}{(z+.2)(z-1)} + \frac{z}{(z+.2)(z-1)^2}$$

$$y_{total} \quad = \quad y_{zi} \quad + \quad y_{zs}$$

Note that the first term is due to the initial conditions, so it is the "zero-input" response. The second part is due to the forcing function and is the "zero-state" response. We can combine them

$$Y(z) = \frac{z^2}{(z+.2)(z-1)} + \frac{z}{(z+.2)(z-1)^2}$$

$$= \frac{z^2(z-1) + z}{(z+.2)(z-1)^2} = \frac{z^3 - z^2 + z}{(z+.2)(z-1)^2}$$

If we do it this way, we are going to wind up taking some nasty derivatives to get the partial fraction expansion of the squared term. It is best to divide the problem in two pieces:

$$\frac{Y_1(z)}{z} = \frac{z}{(z+.2)(z-1)} = \frac{A}{z+.2} + \frac{B}{z-1}$$

In Example A, we had already solved for A and B and gotten A = 0.167 and B = 0.833.

$$\frac{Y_2(z)}{z} = \frac{1}{(z+.2)(z-1)^2} = \frac{C}{z+.2} + \frac{D}{(z-1)^2} + \frac{E}{(z-1)}$$

$$C = \frac{1}{(z-1)^2}\bigg|_{z=-.2} = \frac{1}{(-.8)^2} = 1.56$$

$$D = \frac{1}{(z+.2)}\bigg|_{z=1} = .833$$

$$E = \frac{d}{dz}\frac{1}{(z+.2)}\bigg|_{z=1} = \frac{-1}{(z+.2)^2}\bigg|_{z=1} = \frac{-1}{(1.2)^2} = -.694$$

Now we can group together similar terms

$$Y(z) = \frac{A+C}{z+.2} + \frac{D}{(z-1)^2} + \frac{B+E}{(z-1)}$$

$$= \frac{.833+1.56}{z+.2} + \frac{0.833}{(z-1)^2} + \frac{.167-0.694}{(z-1)}$$

$$y[k] = \left(2.4(.2)^k - 0.527 + 0.833k\right)u[k]$$

Example 5.30. Find the impulse response of the following discrete time linear system:

$$y[k+2] + \frac{1}{6}y[k+1] - \frac{1}{6}y[k] = f[k+1] + f[k].$$

Solution

$$H'(z) = \frac{z+1}{z^2 + \frac{1}{6}z - \frac{1}{6}} = \frac{z+1/2}{(z+1/2)(z-1/3)} + \frac{1/2}{(z+1/2)(z-1/3)}$$

$$= \frac{1}{z-1/3} + \frac{A}{z+1/2} + \frac{B}{z-1/3}$$

$$A = \frac{1/2}{z-1/3}\bigg|_{z=-1/2} = \frac{1/2}{-5/6} = -\frac{3}{5}, \quad B = \frac{1/2}{z+1/2}\bigg|_{z=1/3} = \frac{1/2}{5/6} = \frac{3}{8}$$

$$H'(z) = \frac{1}{z-1/3} + \frac{3/5}{z-1/3} - \frac{-3/5}{z+1/2} = \frac{8/5}{z-1/3} - \frac{-3/5}{z+1/2}$$

Since we did not have a z to divide out and save, we shift,

$$h[k] = \{1.6(1/3)^{k-1} - 0.6(-1/2)^{k-1}\}u[k-1].$$

Example 5.31. Solve the following set of equations for $y_1[k]$:

$$y_1[k+1] - .25y_1[k] = y_2[k-1],$$
$$y_2[k+1] + .5y_2[k] = f[k],$$
$$f[k] = (.5)^k u[k] \quad y_1[0] = 0, \quad y_1[1] = 0.$$

Solution

$$zY_1(z) - zy_1[0] - 0.25Y_1(z) = z^{-1}Y_2(z) + y_2[-1]$$

$$zY_2(z) - zy_2[0] + .5Y_2(z) = \frac{z}{z - .5}$$

We need $y_2[-1], y_2[0]$.

Use the first equation at k = 0:

$$y_1[1] - .25y_1[0] = y_2[-1] \Rightarrow y_2[-1] = 0,$$

and the second equation at k = -1:

$$y_2[0] + .5y_2[-1] = f[-1] = 0 \Rightarrow y_2[0] = 0.$$

$$Y_1(z)(z - 0.25) = z^{-1}Y_2(z)$$

$$Y_2(z)(z + 0.5) = \frac{z}{z - .5}$$

$$Y_2(z) = \frac{z}{(z+0.5)(z-0.5)}$$

$$Y_1(z) = \frac{1}{(z-0.25)(z+0.5)(z-0.5)}$$

235

$$Y_1(z) = \frac{1}{(z-0.25)(z+0.5)(z-0.5)}$$

$$= \frac{A}{(z-0.25)} + \frac{B}{(z+0.5)} + \frac{C}{(z-0.5)}$$

$$A = \frac{1}{(z+0.5)(z-0.5)}\bigg|_{z=.25} = \frac{1}{(.75)(-.25)} = -5.33$$

$$B = \frac{1}{(z-0.25)(z-0.5)}\bigg|_{z=-.5} = \frac{1}{(-.25)(-1)} = 4$$

$$C = \frac{1}{(z-0.25)(z+0.5)}\bigg|_{z=.5} = \frac{1}{(.25)(1)} = 4$$

$$y_1[k] = \{5.33(0.25)^{k-1} + 4(-0.5)^{k-1} + 4(0.5)^{k-1}\}u[k-1]$$

5.6. Stability Analysis in the Z Domain

If we have a discrete system described by the difference equation

$$y[k+2] + a_1 y[k+1] + a_2 y[k] = f[k+1] + b_1 f[k],$$

the Z domain solution will be

$$Y(z) = \frac{(z+b_1)F(z)}{z^2 + a_1 z + a_2} + \frac{I.C.}{z^2 + a_1 z + a_2}.$$

Once again, the behavior of the system will be dictated primarily by the denominator, which is the characteristic equation

$$z^2 + a_1 z + a_2 = (z - p_1)(z - p_2) = 0.$$

This characteristic equation has the roots p_1, p_2. If the two roots are unique and real, the solution will be

$$y[k] = \left[A(p_1)^k + B(p_2)^k\right]u[k].$$

This solution will "blow up" if either p_1 or p_2 has a magnitude greater than one. If they are a complex conjugate pair, the first one will be

$$p_1 = \alpha + j\omega = |p|\angle p_1,$$

and p_2 will be the complex conjugate. The solution will be of the form

$$y[k] = \left[A|p|^k \sin(\angle p_1 \cdot k) + B|p|^k \cos(\angle p \cdot k) \right] u[k].$$

This will blow up if $|p| > 1$.

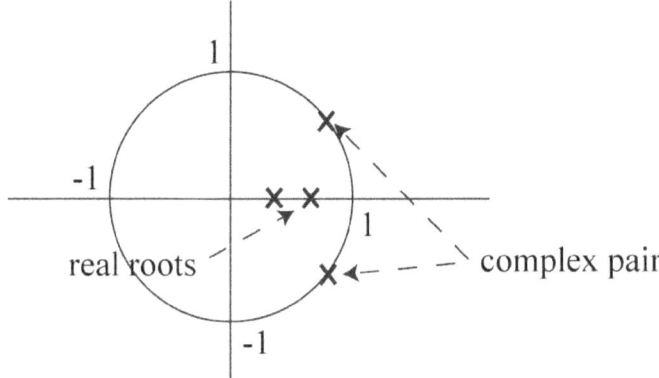

Figure 5.11. The unit circle. If all the roots of the characteristic equation are inside the unit circle, the system is stable. If it has one root on the unit circle, or one pair of complex roots on the unit circle, the system is conditionally stable.

What if we have a root *on* the unit circle? A single root must be 1 or -1, so that means we have a term $\dfrac{z}{z-1}$ or $\dfrac{z}{z+1}$. In the time domain, we will have a term $(1)^k u[k]$ or $(-1)^k u[k]$.
This is called marginally stable. What if we have more than one at a single point, say

$$\frac{z}{(z-1)^2}?$$

This inverse of this is

$$ku[k],$$

which clearly is *not* stable.

5.7. Block Diagrams in the Z Domain

Writing block diagrams for Z domain transfer functions is

similar to block diagrams in the Laplace domain, but there can be differences.

Example 5.32. Find the overall transfer function for the block diagram of Fig. 5.12.

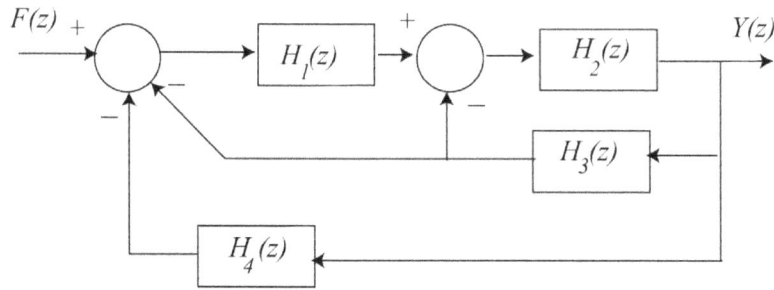

Figure 5.12. A Z-domain transfer block diagram consisting of several individual Z-domain transfer functions.

Solution

The loop in the upper right corner is
$$\frac{H_2}{1+H_2 H_3}.$$
Now the total open loop is
$$\frac{H_1 H_2}{1+H_2 H_3},$$
and the feedback is the parallel addition of $H_3 + H_4$. So
$$H_T = \frac{\dfrac{H_1 H_2}{1+H_2 H_3}}{1+(H_3+H_4)\dfrac{H_1 H_2}{1+H_2 H_3}} = \frac{H_1 H_2}{1+H_2 H_3+(H_3+H_4)H_1 H_2}$$

The block diagram in Fig. 5.11 consisted of negative feedback loops and looked exactly like the block diagrams of Chapter Three. However, Z domain block diagrams can have a significant difference, as seen in the next example.

Example 5.33. Find the transfer function for the positive feedback circuit of Fig. 5.13.

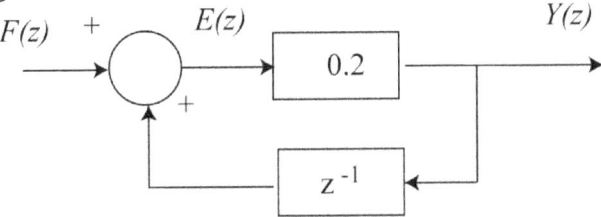

Figure 5.13. The block diagram of a simple digital low pass filter.

Solution

$$E(z) = F(z) + z^{-1}Y(z)$$
$$Y(z) = 0.2E(z)$$

Combining

$$E(z) = F(z) + z^{-1}0.2E(z)$$
$$E(z) = \frac{F(z)}{1 - 0.2z^{-1}};$$

And finally

$$H(z) = \frac{Y(z)}{F(z)} = \frac{0.2}{1 - 0.2z^{-1}} = \frac{0.2z}{z - 0.2}.$$

The inverse Z transform is,

$$Z^{-1}\left\{\frac{0.2z}{z - 0.2}\right\} = 0.2(0.2)^k u[k],$$

which is clearly a stable impulse response, despite the fact that Fig. 5.12 contains positive feedback. Therefore, we cannot use the feedback loop equivalent transfer function, that we had developed for the Laplace block diagrams.

What does the block with z^{-1} represent?

Example 5.34. Find the transfer function corresponding to the block diagram in Fig. 5.14.

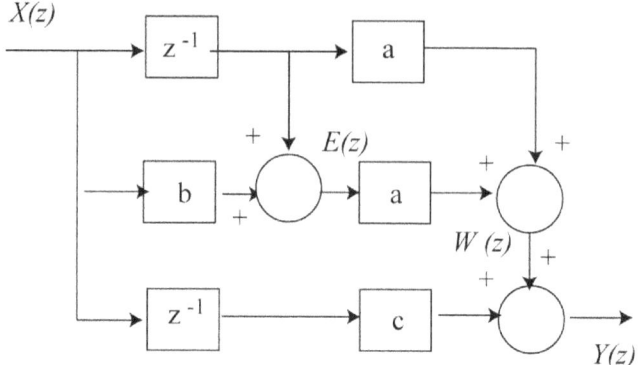

Figure 5.14. A block diagram with no feedback.

Solution

In this case it is better to define an intermediate value after each summation.

$$E(z) = (z^{-1} + b)X(z)$$
$$W(z) = az^{-1}X(z) + aE(z)$$
$$= az^{-1}X(z) + a(z^{-1} + b)X(z)$$
$$= (2az^{-1} + ab)X(z)$$
$$Y(z) = W(z) + cz^{-1}X(z)$$
$$= (2az^{-1} + ab)X(z) + cz^{-1}X(z)$$
$$= ((2a+c)z^{-1} + ab)X(z)$$
$$\frac{Y(z)}{X(z)} = ((2a+c)z^{-1} + ab)$$

What is the impulse response of this circuit?

$$h[k] = ab \cdot \delta[k] + (2a+c)\delta[k-1]$$

If $a = b = c = 1$, what comes out for the following input?

$$x[n] = \begin{cases} 1/2 & n = 0 \\ 1 & n = 1 \\ 1/2 & n = 2 \end{cases}$$

$$X(z) = 0.5 + z^{-1} + 0.5z^{-2}$$

$$W(z) = (1+3z^{-1})(0.5+z^{-1}+0.5z^{-2})$$
$$= 0.5 + 2.5z^{-1} + 3.5z^{-2} + 1.5z^{-3}$$

Example 5.35. Find the transfer function for the block diagram in Fig. 5.15:

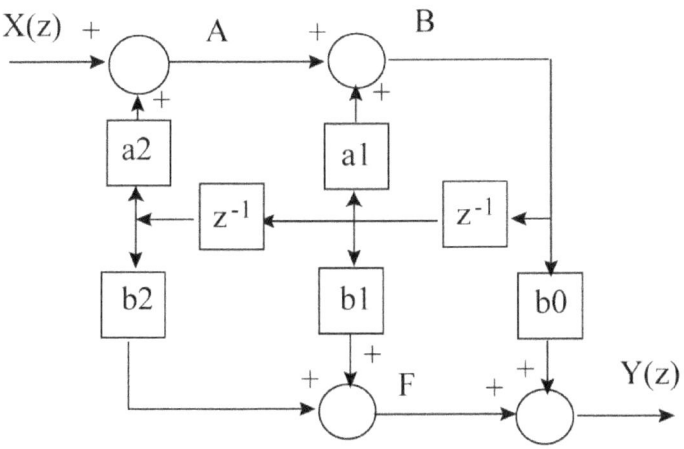

Figure 5.15. A block diagram with feedback.

Solution

First label a new parameter after each summation.
$$A = X + a_2 z^{-2} B$$
$$B = A + a_1 z^{-1} B$$
$$F = b_1 z^{-1} B + b_2 z^{-2} B = (b_1 z^{-1} + b_2 z^{-2}) B$$
$$Y = F + b_0 B = (b_0 + b_1 z^{-1} + b_2 z^{-2}) B$$

Eliminate B:
$$B = (X + a_2 z^{-2} B) + a_1 z^{-1} B$$
$$B = \frac{X}{1 - a_1 z^{-1} - a_2 z^{-2}}$$

Then Y is

$$Y = \frac{b_0 z^2 + b_1 z^{-1} + b_2 z^{-2}}{1 - a_1 z^{-1} - a_2 z^{-2}} X$$

$$= \frac{b_0 z^2 + b_1 z^1 + b_2}{z^2 - a_1 z - a_2} X$$

Example 5.36. What is the step response of the following circuit?

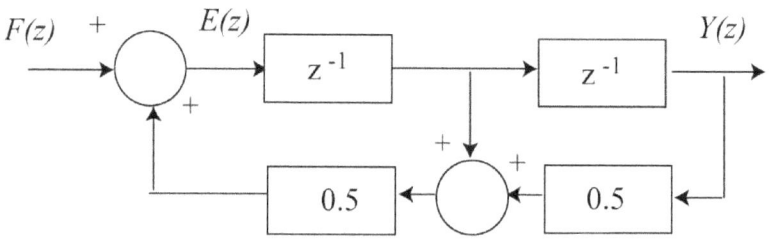

Figure 5.16.

Solution

$$E = F + 0.5 z^{-1} E + 0.25 z^{-2} E$$

$$E = \frac{F}{1 - \frac{1}{2} z^{-1} - \frac{1}{4} z^{-2}}$$

Now notice that

$$Y = z^{-2} E,$$

so

$$Y = \frac{z^{-2} F}{1 - \frac{1}{2} z^{-1} - \frac{1}{4} z^{-2}} = \frac{F}{z^2 - \frac{1}{2} z - \frac{1}{4}}.$$

The step response is

$$Y = \frac{1}{z^2 - \frac{1}{2} z - \frac{1}{4}} \cdot \frac{z}{z - 1}$$

$$p = \frac{0.5 \pm \sqrt{.25 + 1}}{2} = \frac{1}{4} \pm \frac{\sqrt{1.25}}{4} = .25 \pm .559$$

$$p_1 = 0.809, \quad p_2 = -0.309$$

$$\frac{Y}{z} = \frac{A}{z - 1} + \frac{B}{z - 0.809} + \frac{C}{z + .309}$$

242

$$A = \frac{1}{z^2 - 0.5z - 0.25}\bigg|_{z=1} = \frac{1}{.25} = 4$$

$$B = \frac{1}{z-1}\frac{1}{z+.309}\bigg|_{z=.809} = -0.214^*$$

$$C = \frac{1}{z-1}\frac{1}{z-.809}\bigg|_{z=-0.309} = 1.46$$

$$Y(z) = \frac{4z}{z-1} + \frac{-0.214z}{z-0.809} + \frac{1.46z}{z+.309}$$

$$y_{step}[k] = \left(4 - 0.214(0.809)^k + 1.46(-.309)^k\right)u[k]$$

Example 5.37. Find the transfer function of the block diagram in Fig. 5.17

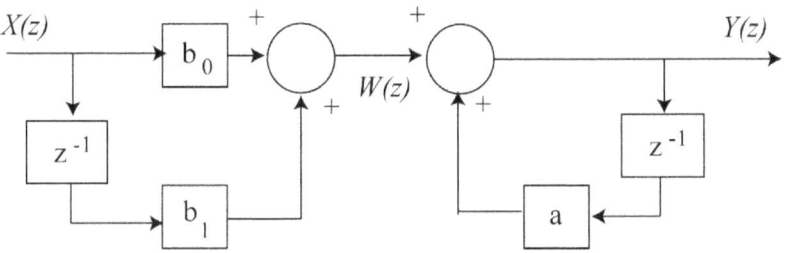

Figure 5.17. A canonical structure.

Solution

$$W = (b_0 + b_1 z^{-1})X$$
$$Y = W - az^{-1}Y$$
$$Y = \frac{W}{1 + az^{-1}}$$
$$H = \frac{Y}{X} = \frac{b_0 + b_1 z^{-1}}{1 + az^{-1}}$$

The structure in Fig. 5.6 is considered to be one of the canonical structures of digital filtering. It can be extended in the following manner

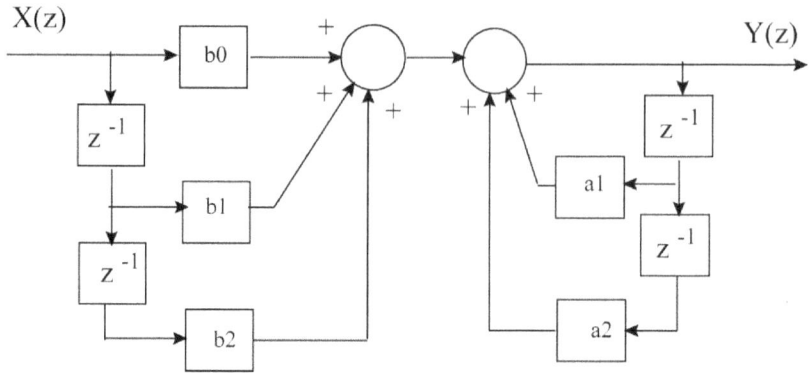

Figure 5.18. A canonical digital filter structure similar to Fig. 5.17, but one order higher.

The transfer function is now

$$H = \frac{Y}{X} = \frac{b_0 + b_1 z^{-1} + b_2 z^{-2}}{1 + a_1 z^{-1} + a_2 z^{-2}}.$$

Example 5.38. Find the transfer function corresponding to the block diagram

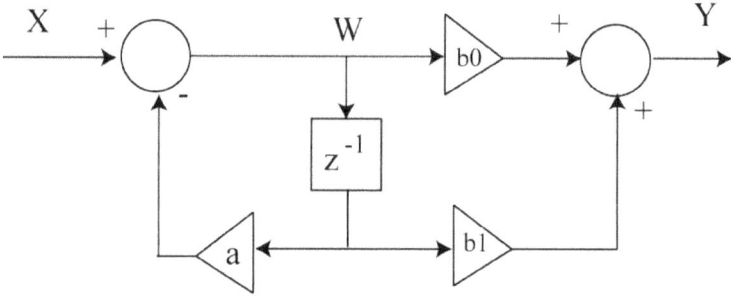

Figure 5.19.

Solution

$$W = X - az^{-1}W$$
$$W = \frac{X}{1 + az^{-1}}$$
$$Y = b_0 W + b_1 z^{-1} W$$

The two together give

244

$$\frac{Y}{X} = \frac{b_0 + b_1 z^{-1}}{1 + az^{-1}}.$$

Example 5.39. Find the transfer function corresponding to the following block diagram

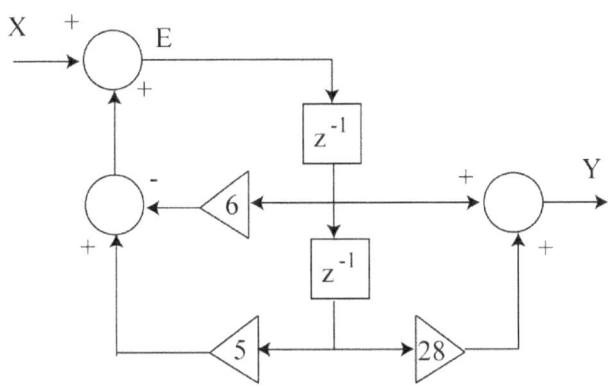

Figure 5.20.

Solution

$$E = X - 6z^{-1}E + 5z^{-2}E$$

$$E = \frac{X}{1 + 6z^{-1} - 5z^{-2}}$$

$$Y = z^{-1}E + 28z^{-2}E$$

$$\frac{Y}{X} = \frac{z^{-1} + 28z^{-2}}{1 + 6z^{-1} - 5z^{-2}}.$$

Example 5.40 Consider the block diagram below. For what values of a is the system stable?

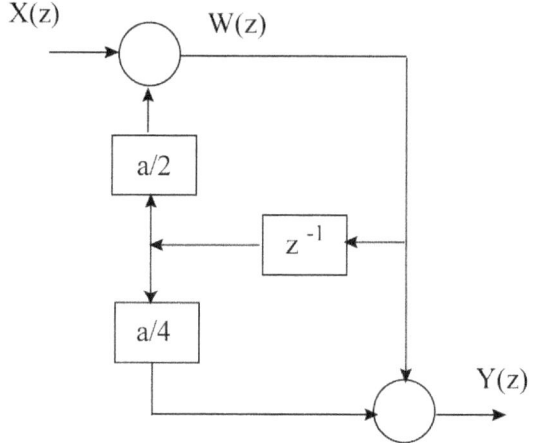

Figure 5.21.

Solution

$$W = X + \frac{a}{2}z^{-1}W$$

$$W = \frac{1}{1-(a/2)z^{-1}}X$$

$$Y = W + \frac{k}{4}z^{-1}W = \left(1 + \frac{k}{4}z^{-1}\right)W$$

$$\frac{Y}{X} = \frac{1-(a/4)z^{-1}}{1-(a/2)z^{-1}} = \frac{z+(a/4)}{z-(a/2)}.$$

It is stable for

$$|a| < 2.$$

Table 5.1 Some common Z Transform pairs

1.	$\delta[k]$	1
2.	$u[k]$	$\dfrac{z}{z-1}$
3.	$a^k u[k]$	$\dfrac{z}{z-a}$
4.	$k a^k u[k]$	$\dfrac{az}{(z-a)^2}$
5.	$k^2 a^k u[k]$	$\dfrac{az(z+1)}{(z-a)^3}$
6.	$a^k \cos(\omega T k) u[k]$	$\dfrac{z^2 - a\cos(\omega T)z}{z^2 - 2a\cos(\omega T)z + a^2}$
7.	$a^k \sin(\omega T k) u[k]$	$\dfrac{a\sin(\omega T)z}{z^2 - 2a\cos(\omega T)z + a^2}$

Table 5.2. Z Transform Properties

1. $Z\{\alpha f_1[k] + \beta f_2[k]\}$ $\alpha F_1(z) + \beta F_2(z)$
2. $Z\{f[k-n_0]u[k-n_0]\}$ $z^{-n_0}F(z)$
3. $Z\{f[k-1]u[k]\}$ $z^{-1}F(z) + f[-1]$
4. $Z\{f[k-2]u[k]\}$ $z^{-2}F(z) + z^{-1}f[-1] + f[-2]$
5. $Z\{f[k+1]u[k]\}$ $zF(z) - zf[0]$
6. $Z\{f[k+2]u[k]\}$ $z^2F(z) - z^2f[0] - zf[1]$
7. $Z\{kf[k]\}$ $-z\dfrac{d}{dz}F(z)$
8. $Z\{a^k f[k]\}$ $F\left(\dfrac{z}{a}\right)$
9. $Z\{f[k] * g[k]\}$ $F(z)G(z)$
10. $\lim_{k \to 0} f[k]$ $\lim_{k \to \infty} F(z)$
11. $\lim_{k \to \infty} f[k]$ $\lim_{k \to 1}\left\{\dfrac{z-1}{z}F(z)\right\}$

References

1. Z. Gajic, "*Linear Dynamic Systems and Signals,* Upper Saddle River, NJ: Prentice Hall, 2003.

2. M. J. Roberts, *Signals and Systems—Analysis Using Transform Methods and MATLAB, 2nd Ed*, New York, NY: McGraw Hill, 2012.

Problems

5.1.1. Graphically convolve the following two signals. Each graph is the finite length that we see.

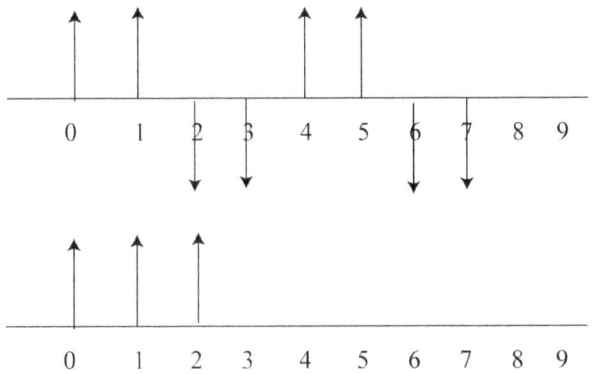

5.1.2 Two function are given by

$$f_1[k] = \begin{cases} 1 & k = 0 \\ 3 & k = 2 \\ 0 & \text{otherwise} \end{cases} \quad f_2[k] = \begin{cases} 2 & k = 0 \\ 1 & k = 3 \\ 0 & \text{otherwise} \end{cases}$$

What is $f_1 * f_2$, i.e., the convolution of the two?

5.1.3 A discrete system is described by the following difference equation:
$$y[k+2] + 0.4y[k+1] + 0.1y[k] = x[k]$$
Write the first four terms of the impulse response.

5.2.1. Find the Z transforms of the following:

a. $f_1[k] = \sin(45°k)(u[k] - u[k-2])$

b. $f_2[k] = (.3)^k \sin(30°k) u[k]$

c. $f_2[k] = e^{-k-2} \sin\left(\frac{\pi}{2}k\right) u[k]$

5.2.2 Using the Z transform of $a^k u[k]$ and the Euler formulas, find the Z transforms of:

a. $\sin(\omega Tk)$

b. $a^k \sin(\omega Tk)$

5.2.3. (Problem 5.6 in Gajic). Given $f[k] = ku[k]$, find the Z transform of the following signal

$$(k-1)f[k] + 5^k f[k] + f[k-1]u[k-1] + 5f[k+3]u[k].$$

5.3.1. Find the inverse Z transforms of the following functions:

a). $F_1(z) = \dfrac{z+1}{(z-1)^2}$

b). $F_2(z) = \dfrac{1}{(z-.2)^2}$

c). $F_3(z) = z^{-2}(z^{-1} + z^{-2})$

d). $F_4(z) = \dfrac{1}{z^2 + 2z - 5}$

e) $F_5(z) = \dfrac{1}{(z-1/2)(z-1/3)}$

5.3.2 Solve for y[k] for the given forcing functions. Assume all initial conditions are zero.

$$y[k] - \frac{3}{4}y[k-1] + \frac{1}{8}y[k-2] = f[k-2].$$

Check the first three terms of your solution against the first three terms obtained directly from the equation.

a. $f[k] = u[k]$

b. $f[k] = (0.5)^k u[k]$

5.3.3 Find the inverse Z transform of the following function. Write your answer in the most concise form possible.

$$H(z) = \frac{z-1}{z^2 - 1.6z + .64}.$$

5.4.1. Find the inverse Z transforms of the following:

a). $F_1(z) = \dfrac{z}{z^2 + 2z + 5}$

b). $F_2(z) = \dfrac{1}{z^2 - z + 0.5}$

c). $F_3(z) = \dfrac{z^2 + 2z + 2}{z^2 - 2z + 4}$

d). $F_4(z) = \dfrac{5}{z^2 - .4z + 0.08}$

e). $F_5(z) = \dfrac{1}{z(z^2 + 2z - 3)}$

5.4.2 Solve for y[k]. Assume all initial conditions are zero.

$$y[k+2] - \frac{1}{2}y[k+1] + \frac{1}{8}y[k] = u[k]$$

5.5.1 Solve for the Y(z) that would lead to the correct y[k], i.e., you don't have to actually get the inverse Y(z).

$$y[k+1] + \frac{5}{6}y[k] + \frac{1}{6}y[k-1] = f[k]$$

$$y[0] = \frac{6}{5}, \quad y[1] = 0, \quad f[k] = \left(\frac{1}{3}\right)^k u[k]$$

5.5.2 Find the response of the system described by the following difference equation:

$$y[k] + 0.7y[k-1] - 0.3y[k-2] = f[k],$$
$$y[0] = 1, \; y[-1] = 1, \quad f[k] = \delta[k].$$

5.5.3 A system is described by the difference equation

$$y[k+1] - 2y[k] + 3[k-1] = f[k]$$

with initial conditions $y[0] = 1$, $y[1] = 0$. Find the complete response if $f[k] = \delta[k]$

5.5.4 Solve for y[k].

$$y[k+2] + 0.6y[k+1] - .1y[k] = f[k+1]$$
$$y[-1] = 0, \quad y[0] = 0, \quad f[k] = \delta[k]$$

5.5.5 Solve for y[k]:

$$y[k+2] - 0.4y[k+1] - 0.45y[k] = 0, \qquad y[-1] = y[-2] = 1.$$

5.5.6. Solve the following for y[k]:

$$y[k+1] - \left(\frac{3}{4}\right)y[k] + \frac{1}{8}y[k-1] = (1/4)^k u[k]$$
$$y[0] = 0, \quad y[-1] = 0.$$

5.6.1 Determine if the system described by the following difference equation is stable or not. Give the reason.

$$y[k+1] - 0.5y[k] + y[k-1] = f[k+2] + 3f[k].$$

5.6.2 Determine if the system described by the following difference equation is stable or not. Give the reason.

$$y[k+2] - y[k+1] + \frac{1}{4}y[k] = f[k+1].$$

5.7.1. Find the transfer function corresponding to the block diagram below.

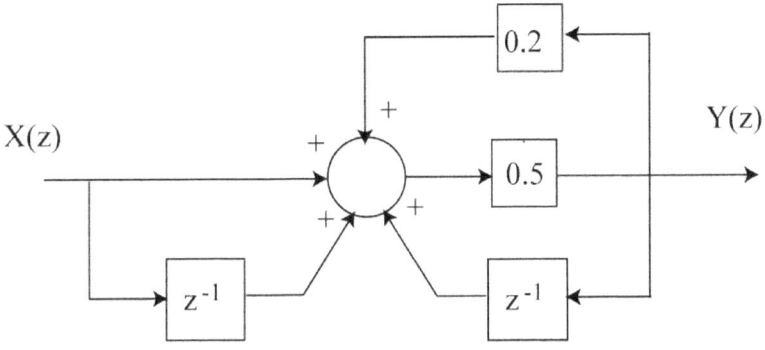

5.7.2 Find the transfer function corresponding to the following block diagram

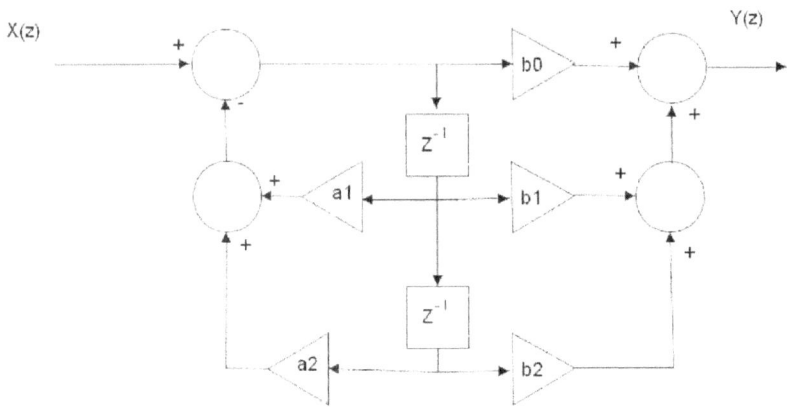

5.7.3 Find the transfer function of the following block diagram.

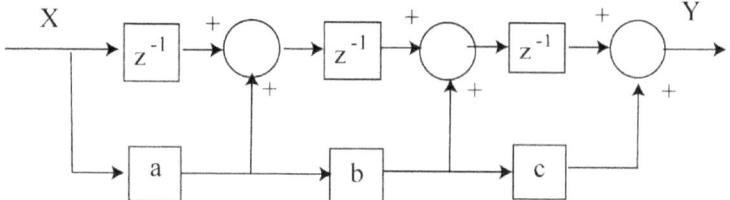

5.7.4 Find the transfer function of the following block diagram.

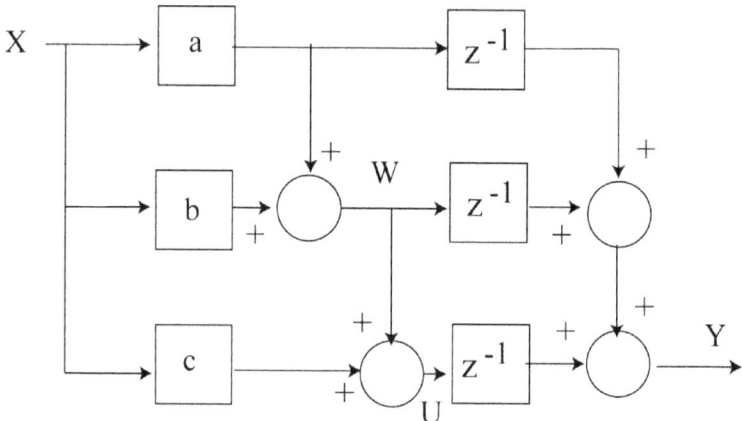

5.7.5 Find the transfer function of the following block diagram.

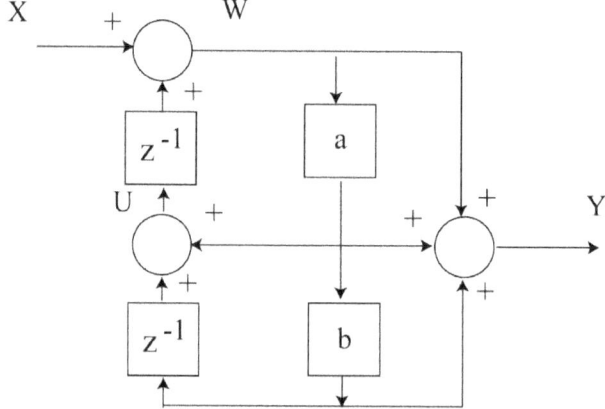

5.7.6 Determine the transfer function of the following block diagram

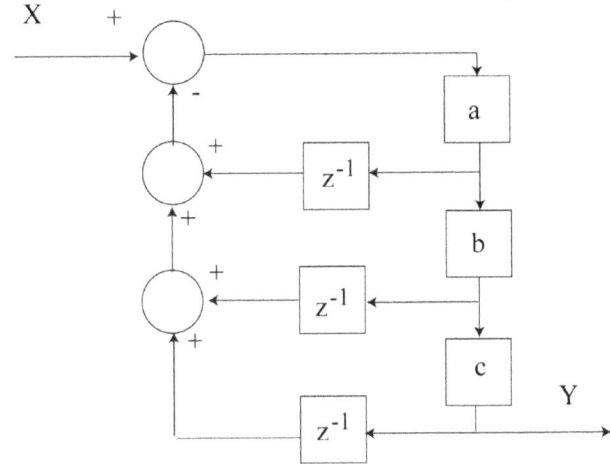

5.7.7 Find the transfer function of the following block diagram.

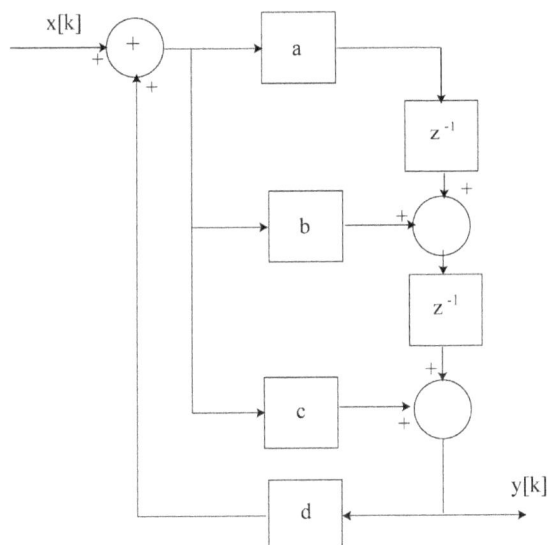

www.ingramcontent.com/pod-product-compliance
Lightning Source LLC
Chambersburg PA
CBHW071019240526
45469CB00006BD/1991